P
Dy

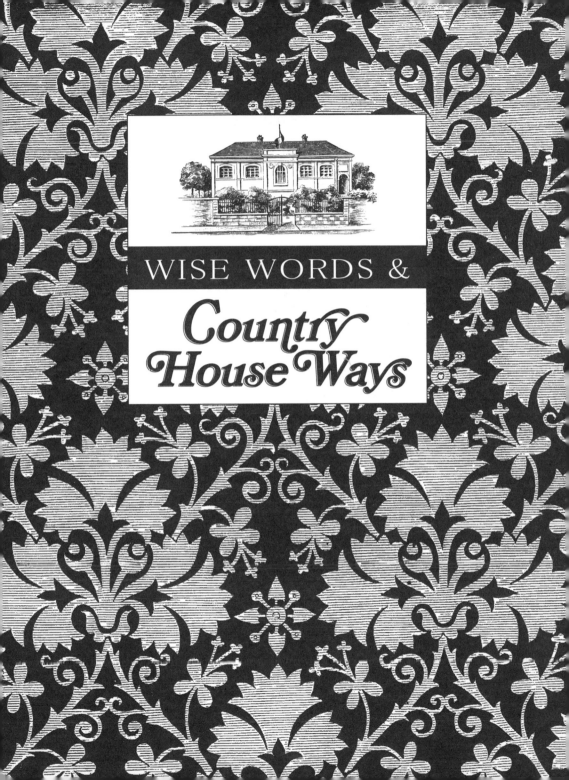

WISE WORDS &
Country House Ways

To Ursula
for all her encouragement

Ruth Binney has been collecting old sayings, wisdom and traditional remedies relating to country matters of all kinds for over 50 years. She holds a degree in Natural Sciences from Cambridge University and has been involved in countless publications during her career as an editor. She is the author of seven other books for David &Charles.
www.ruthbinney.com

RUTH BINNEY

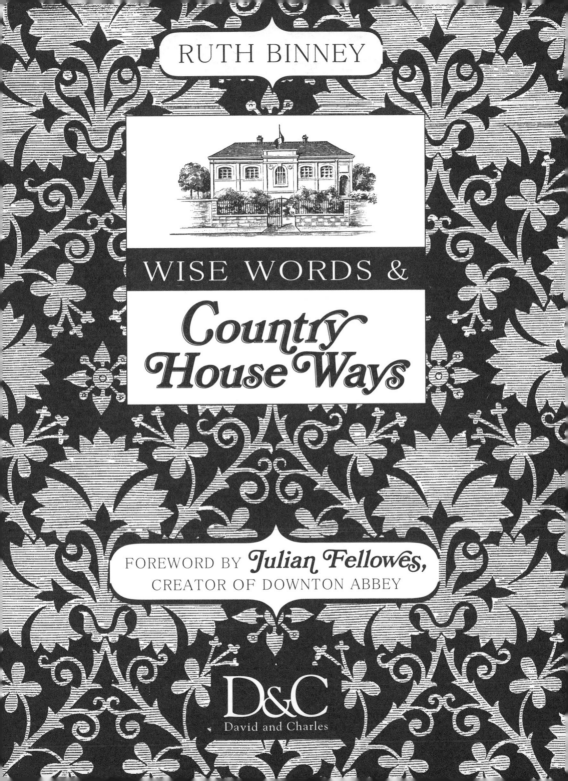

WISE WORDS &

Country House Ways

FOREWORD BY *Julian Fellowes,*
CREATOR OF DOWNTON ABBEY

D&C
David and Charles

CONTENTS

FOREWORD BY JULIAN FELLOWES   5

INTRODUCTION   6

CHAPTER 1 KEEPING HOUSE   8

CHAPTER 2 THE DAILY ROUTINE   40

CHAPTER 3 THE COUNTRY HOUSE KITCHEN   76

CHAPTER 4 A MATTER OF MANNERS   106

CHAPTER 5 ENTERTAINMENT, LEISURE AND SPORT   136

CHAPTER 6 GARDEN AND GROUNDS   168

INDEX   205

Foreword

Other nations may have supportable claims to lead in the fields of food or fashion or modern technology, but no one could seriously challenge the British when it comes to life in a country house. The great house, with its gardens and dependencies, with its park and farmland, its stables and carpentry shops and kitchens and fisheries, was a kingdom in itself, and if not quite self-supporting, it was often almost so. Perhaps it is this sense of completeness, a world sufficient unto itself, that explains the continuing fascination with all that the English country house represents.

It was an arrangement that involved men and women from every background who each had a crucial role to play in this complex machine. In fact, one of the most pleasing of the recent developments in our appreciation of the country house is that we have begun to see that these places were part of our shared history and not simply the homes of the upper classes. When I was young there was a tendency, with any house in the care of the National Trust or opened by its owners, to put only the public rooms on display for the ticket buyers to marvel at. And they were very impressive, those drawing rooms, libraries and gilded ballrooms, but there was no sign of the kitchens or work places, thereby denying the visitor any real sense of how it worked. One of the first houses to break with this rule was Llanhydrock, in Cornwall. A fire in the 1880s had destroyed much of the original and the rebuild had resulted in a series of reception rooms designed according to the not very inspiring standards of the late 19th century, but, by contrast, the state-of-the-art kitchens, the pantries, the brushing and ironing and lamp rooms could not have been more interesting. The National Trust then commendably decided to make it a priority to open these places and, for once, to give the public a real idea of the bee-hive nature of a great house, where the work rooms were intimately linked to the lives of the family in a continuous cycle of labour, in which no detail was too small not to be governed by rules and custom and etiquette and costume.

This is the world which Ruth Binney has brought so wonderfully to life in her book. To a modern eye, there was certainly a good deal of unfairness involved in the arrangement, there is no point in pretending otherwise, but there was also a level of dedication and skill and, above all, commitment, from both groups, the family and their servants, that seems somehow admirable, especially when viewed from our rather selfish and undisciplined century. I do not believe we should be too nostalgic for the ways that are gone, but, by the same token, nor do I believe we should be too afraid of learning from their example.

Julian Fellowes, creator of Downton Abbey

Introduction

There is no doubt that curiosity about the British country house, the way it was run, the families and staff who lived in it – and the guests they entertained – has been greatly stimulated by award-winning series such as the hugely popular Downton Abbey. *Now* Wise Words and Country House Ways, *which concentrates largely on life in the Georgian, Victorian and Edwardian eras, reveals entertaining and informative insights into the detailed life and workings of the country house, starting from a selection of the hundreds of contemporary maxims used to advise and instruct the members of the household and their guests. These range from the details of managing the home and garden to entertaining visitors – including royalty – to the duties of the many and varied members of staff.*

The size and splendour of a country house and its estate were a reflection of the status and ambition of its owner, and the whole was a world in itself, which if not totally self-sufficient was often nearly so. The intricate way in which the country house worked is reflected in the six chapters of the book, beginning with 'Keeping House' and progressing to 'The Daily Routine' and 'The Country House Kitchen'. Since correct behaviour was so important to all activities of the house, 'A Matter of Manners' addresses the essentials of etiquette, a theme that also extends into 'Leisure, Entertainment and Sport', whether this was hunting, fishing or attending a ball or country house wedding. Finally, 'Garden and Grounds' focuses on everything from the cultivation of exotic fruit for the table to brewing and the care of horses and other animals.

Whenever possible, information for the book has come from contemporary sources, notably The Complete Servant *of 1675*, The Servants' Practical Guide *of 1880 (subtitled 'A handbook of duties and rules', and written by the author of 'Manners and Tone of Good Society'),* Our Homes and How to Make Them Healthy *of 1883, edited by Shirley Forster Murphy, the Victorian manual* Enquire Within *of 1894, and the four volumes of* Cassell's Household Guide *of the 1880s. On kitchen and household matters, Mrs Beeton had much pertinent advice worth including, while for the garden William Robinson's* The English Flower Garden *(1883) and Robert Thompson's* The Gardener's Assistant *(1859) proved invaluable. Any contemporary 'recipes' included, such as those for cleaning materials, have not been tested – they are here for interest and their use is at the reader's own risk.*

Without the resources of the London Library, and the many volumes lent by friends and family, the intriguing experience of researching and writing this book would have been impossible, and I thank them all for their help and support. I am also indebted to the team at David & Charles for their care and enthusiasm in the book's creation and of course to Julian Fellowes for his foreword.

Ruth Binney

West Stafford, Dorset, May 2012

CHAPTER 1
KEEPING HOUSE

W hether it was bought or inherited, managing a country house was no mean task. Keeping the house running well demanded not only considerable resources but also the managerial skills required to hire, employ and keep happy a veritable army of staff. The house needed to be properly furnished and decorated, kept up to date with the latest advances, and kept warm, comfortable and clean for everyone living in it, as dictated by the standards of the day. The finances of a well-run house were always a top priority. Not only were precise records of all expenditure mandatory, but they needed to be examined on a regular basis to make sure that no fraud of any kind was being committed.

The way in which the typical country house was organized and arranged followed a definite pattern, with an impressive entrance hall, drawing and dining rooms, and staff quarters well separated from those of the family and ostensibly invisible. The master and mistress of the house would have their own 'apartments', often with separate bedrooms, and rooms – or suites – set aside for key family members and for important visitors, who might include royalty. Other visitors would each be offered their own rooms, while staff were often accommodated in shared rooms or dormitories.

The arrival of flushing lavatories and bathrooms with running water did a great deal to improve the comfort of the country house, which, in the days before central heating, would have open fires in all family rooms in use during the winter, or on any day deemed cold enough to make them necessary. Equally, the installation of electricity, although it caused initial anxieties, revolutionized home lighting and allowed for the use of gadgets, while the telephone provided speedy contact with the outside world.

THE HEAD OF THE HOUSEHOLD IS MASTER OF HIS REALM

This entailed care of both the house and estate, which, with the help of his family and servants, he needed to run as efficiently as possible.

Fair treatment was the obligation of the master to his servants. This included rewarding them for good service, but he was also expected to ensure that they were well fed and housed, to have regard for their health and, if necessary, to intervene on their behalf if they ran into trouble of any kind. And when long-term, elderly servants reached retirement age he was expected to provide them with an annuity of some kind to support them for the rest of their lives.

As head of the family, the master of the country house was entitled to the obedience of everyone in it, including his wife and children. In return for their employment he demanded absolute loyalty from his servants, down to the lowliest scullery maid or hall boy. He could chastise, encourage or reward them according to their performance – and his own whims. But of all the virtues required of his servants, loyalty and discretion were the most important. Without them there was a grave risk of dishonour being brought on the household.

The master-servant relationship was intended to keep the house and estate running smoothly, but the master of the house could behave exactly as he wished. Recording his visit to Belvoir Castle in Leicestershire in 1838 the memoir writer Charles Greville says: 'The Duke of Rutland is as selfish a man as any of his class – that is, he never does what he does not like, and spends his whole life in a round of such pleasures as suit his taste, but he is neither a foolish nor a bad man, and partly from a sense of duty, partly from inclination, he devotes time and labour to the interest and welfare of the people who live and labour on his estate.'

While the master of the house might also conduct business in London, or serve in the armed forces, his family would remain in the country for most of the year, going 'up to town' only for the London season during the summer months.

Hospitality needed to be the watchword of the master of the country house, whether he was entertaining royalty and the aristocracy or hosting servants' balls at Christmas or other times of the year (see Chapter 5). However, strict discipline was also enforced, with unruly servants often dismissed without warning and without the vital references needed in order to obtain another position. In the 18th century any servant found forging such a testimonial might be sentenced to a year's hard labour.

GOOD TEMPER SHOULD BE CULTIVATED BY EVERY MISTRESS ...

One of the many maxims of Mrs Beeton, and one adhered to by the best mistresses of country establishments to help ensure that all ran smoothly in the household.

Expanding on this sentiment, the advice continues '... as upon it the welfare of the household may be said to turn; indeed, its influence can hardly be over-estimated, as it has the effect of moulding the characters of those around her, and of acting beneficially on the happiness of the domestic circle. Every head of a household should,' she adds, 'strive to be cheerful and should never fail to show a deep interest in all that appertains to the well-being of those who claim protection of her roof.'

Those under such protection and serving within the house, and in places such as the dairy and brewhouse, could – up to the 20th century – number several dozen. Although it was the principal task of the head housekeeper to ensure that all domestic work was completed well and in a timely manner, the ultimate responsibility lay with the mistress of the house, and her attitude could have a great impact. The servants managed directly by her included her own lady's maid, the nurse, the governess and, critically, the cook.

As well as their physical welfare, the good country house mistress would take on responsibility for the morals of her staff. At Holkham Hall in Norfolk, for instance, servants were forbidden to gamble, swear or be seen out of the village after dark. On a lighter note, maids were banned from using curl-papers in their hair (these being regarded as a total frivolity) and discouraged from dressing, when off duty, in any but the plainest clothes. Sunday church attendance was, of course, vigorously encouraged, if not compulsory.

∽ KEEPING STAFF HAPPY ∾

In well-run households the staff were considerately treated and allowed a variety of perks. Typical examples from the records of Hatfield House in the 19th century included:

For each manservant: a pint of beer (home brewed) at lunch, dinner and supper.

For each woman servant: a pint of the same beer at lunch and supper.

Cook: the right to bones and dripping, which could be used or sold.

Lady's maid: her mistress's cast-off clothes and accessories.

Butler: candle ends and empty bottles, free to be sold.

Coachman: wheels from carriages, but allowed only if the servant had been in service for more years than the age of the wheels.

THE OFFICE OF HOUSE STEWARD IS OF SUPREME IMPORTANCE

In large country houses the steward was the head of the staff. He needed to be loyal, trustworthy and a man of manners, au fait with everything from accounting to the nuances of etiquette.

The chief duties of the house steward were the management and hiring of staff (except for personal servants such as valets, ladies' maids and nurses) and the economy of the household. As the 'job description' clearly set out in *The Country Gentleman's Vademecum* of 1699 underlines, the steward must '... receive and pay all Monies, buy in the Provision for the Family, hire all Livery-men, buy all Liveries, pay all Wages, direct and keep in order all Livery-men (except the Coachman and Groom), to be at his Master's Elbow during Dinner, and receive all Orders from him relating to Government; to oversee and direct the Baliff, Gardener, &c. in their Business; and also the Clerk of the Kitchen, Cook, Butler &c. to whom he delivers the Provision, Wine, Beer, &c. who give him an Account of the spending it, weekly or otherwise.'

Keeping accurate double-entry accounts was a vital role of the house steward. As *The Complete Servant* explained, he must: '... keep an account of the monies received by him, on one page, and of monies paid or disbursed by him, on the opposite page; and these two pages being cast up, and the amount of one side being deducted from the amount of the other, will, if the account has been correctly kept, shew at once the exact balance, belonging to his employer, remaining in his hands.' He needed to work with the housekeeper to ensure that all provisions brought into the house were up to quality and the bills for them checked and paid.

The steward had his own room in the house and the dedicated service of one or more steward's room men or boys. The footmen would also attend to his needs.

LIVERY IS PARTICULAR TO EVERY HOUSEHOLD

The 'uniform' of some male servants, livery reflected greatly on the wealth and position of the family in which they were employed.

Livery was worn by the servants most on display, particularly footmen of all ranks. The youngest liveried servants were the 'tigers': small boys who perched behind light carriages to act as grooms. Livery was exclusive to each household, allowing families to pick out their own servants in a crowd. Lord Leicester's men, for instance, wore blue coats, while those of the Marquis of Bath at Longleat wore mustard yellow. Buttons were silver or gold. The traditional pattern of the waistcoat, often trimmed with gold braid, was horizontal stripes. Silk stockings and shoes with silver buckles were *de rigueur*.

For evenings, as well as donning white ties, liveried footmen might be required to powder their hair. To do this they would dip their heads in water, rub in soap to create a lather, comb it through, then apply powder from a powder puff – either a specially prepared coloured powder (violet was commonly preferred) or ordinary flour. They then had to contrive to spend the entire evening in service without letting any powder fall onto their jackets – a feat that involved standing very still for long periods.

All liveries belonged to the master, not the servant. Daytime liveries were usually renewed twice a year, in spring and autumn, evening ones annually. When a servant left the household the livery always had to be returned.

Coachmen also wore livery, though of a different style to that of footmen. Their dress included black boots with white tops, red or other brightly coloured greatcoats resplendent with brass buttons bearing the family crest, and hats with feathered cockades.

A HOUSEKEEPING ACCOUNT BOOK MUST INVARIABLY BE KEPT

An absolute essential of running a country house, and the daily responsibility of the housekeeper in collaboration with the house steward.

The housekeeper needed to keep a meticulous record of the amounts paid out each day, however small. At the end of the month, advised Mrs Beeton, '... let these payments be ranged under their specific heads of Butcher, Baker, &c.; and thus will be seen the proportions paid to each tradesman, and any one month's expenses compared with another ... When, in a large establishment,' she continues, 'a housekeeper is kept, it will be advisable for the mistress to examine her accounts regularly. Then any increase of expenditure which may be apparent, can easily be explained, and the housekeeper will have the satisfaction of knowing whether her efforts to manage her department well and economically have been successful.'

> *Everything coming into the house was checked immediately by the housekeeper for quality and 'to see that in weight and measure they agree with the tickets sent with them'. Only then could they be allocated for use.*

In houses where there was a house steward, the bills would be examined and paid by him. However, the housekeeper still needed to keep a daily eye on the larder, in collaboration with the cook, to ensure that it was fully stocked, particularly when visitors were due. 'The best and most economical way possible for marketing,' said *The Complete Servant* 'is to pay ready money for all that you can, especially for miscellaneous articles, and to deal for the rest with the most respectable tradesmen.' The housekeeper also needed to keep a record of provisions given to the staff, which would include soap, tea, sugar and beer for visitors' servants.

EVERY HALL SHOULD BE DESIGNED TO IMPRESS

The typical country house had halls of two kinds – one for the family and their guests, the other for the servants.

The hall began life in the medieval country house as a space where everyone in the household ate their meals. To mark their distinction the lord and his family would sit on a raised dais – at the 'high table' – with all the other diners, including servants, ranged below them according to status. Originally little more than a covered courtyard, the medieval hall had typically become endowed, by the 14th century, with a magnificent timber roof. Adjacent to the hall were the private chambers of the lord, his wife and family, while adjoining at the opposite end were the kitchens.

As country house architecture developed, so rooms began to be built above the hall, which had originally stretched right up to the roof. Galleries constructed around it were ideal for musicians, who would play during dinner and for dancing and other entertainments. Not until the 17th century did servants begin eating in their own separate hall – the first step in the gradual transformation of the country house hall into a grand entrance that could also be used for games such as billiards, or as a ballroom. Key to this change was

the work of the architect Inigo Jones who, inspired by Italian villas, added the dining saloon to the house.

Pronouncing on the ideal arrangement of the large Victorian house, *Our Homes and How to Make Them Healthy* advised: 'The servants' hall should be placed within easy access of the entrance hall, and also close to the servants' entrance, so that it may be used as a waiting room when occasion demands,' adding, 'The windows should be so contrived that, while they should not overlook the private grounds, they should yet afford a cheerful and pleasant outlook onto the kitchen yard or the garden.' By this time, servants usually had their own quarters, often in a separate wing, 'not only for the comfort and convenience of the family, but also for that of the servants.'

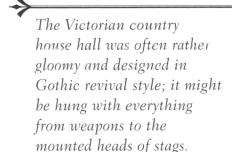

The Victorian country house hall was often rather gloomy and designed in Gothic revival style; it might be hung with everything from weapons to the mounted heads of stags.

⌘ DECORUM IN THE SERVANTS' HALL ⌘

As elsewhere in the house there were strict rules of conduct in the servants' hall, which was said by The Complete Servant to be a little world by itself, in which 'passions, tempers, vices and virtues, are brought into play, and contribute their full share in promoting welfare and happiness':

The housekeeper sits at the head of the table, with the cook on her right and the lady's maid on her left.

The butler sits at the opposite end with the under butler on his right and the coachman on his left.

The dinner is set on the table by the cook, and the beer is drawn by the under butler.

Servants of coarse manners, vulgar habits, or profane discourse and malicious dispositions, are shunned by others and never make good their footing in first-rate families.

THE DRAWING ROOM ALLOWS FOR QUIET AND SEPARATION

That is of ladies from men. As it evolved, the country house drawing room became an increasingly elaborate showcase for fashionable worldly goods.

The drawing room, or properly the 'withdrawing' room, began as an annexe to the bedchamber, to which the most privileged guests would be admitted. Later, it moved downstairs and became a room in which people would assemble before dinner and afterwards, if necessary, wait while the dinner table was being cleared. From this it developed into a public room in which visitors could also be entertained during the afternoon.

Music was customarily played in the drawing room, which would therefore be furnished with a harpsichord – replaced later by the grand piano – and possibly with other instruments such as a harp and a cello.

By the Victorian era the country house drawing room was generally full of furniture, as this description by the garden designer J.C. Loudon illustrates: 'A large round table is usually placed in the middle of the drawing-room, on which are generally books of prints and other things to amuse the company ... Two card tables would stand one on each side of the fireplace: and, besides all these, we must have tables of various sizes, some small ones on pillars; a chess table, with an inlaid marble top, the men placed upon it; a large china dish set in a gilt sort of tripod; a sort of table flower-stand; and I cannot tell what besides ... Writing, work, and drawing boxes of handsome kinds

and everything amusing, curious, or ornamental, is in its place in the drawing-room; but the host of trumpery toys so often seen there would be unworthy of a place in a room like this.'

For comfort, the Regency drawing-room would customarily be furnished with a 'sofa table' – a low table, usually with a flap at each end, set alongside a sofa on which a lady could recline with ease. Rosewood was a favourite material, as it was for other furniture of the period, particularly chairs.

✑ RELAXING IN SPLENDOUR ✑

Some of the many country house drawing rooms of note include:

Ham House, Surrey – The north drawing-room, with a baroque chimneypiece decorated with cherubs and fine tapestries based on paintings of the seasons by the Flemish artist David Teniers.

Syon House, Middlesex – The crimson drawing room, with damask hangings and a ceiling ornamented with squares and octagons each enclosing a painted panel by Angelica Kauffmann. The carpet of red, gold and blue was woven especially for the room.

Saltram House, Devon – Designed by Robert Adam, with a coved ceiling of rose pink and green decorated with circular paintings on sky blue backgrounds, a simple fireplace and eye-catching Adam carpet.

Standen, West Sussex – Designed by Philip Webb, with William Morris wallpapers and textiles.

WALLPAPER IS ONE OF THE MOST IMPORTANT PARTS OF THE ENRICHMENT OF A HOUSE

From the 17th century wallpaper was used to embellish country houses, being designed originally to imitate decorative fabric wall hangings.

Wallpaper was first produced in Britain in small block-printed sheets, rather like tiles, which were used as a substitute for the thick, often dark, woven fabric wall coverings that had served as both decoration and insulation. Many 19th-century wallpapers were printed with pigments derived from substances such as lead and arsenic, and they could be a health hazard. Even in 1895 home owners were warned: 'When having a house papered, make quite sure before they are put on the walls that they are not arsenicated ... if you are doubtful about them submit a piece of each paper to a chemist or analyst and ask his opinion. Many cases of persistent illness have been traced to arsenic in the wall-paper, and it is not only present in green papers, but also in those of other colours.'

Because papers were hand printed, hanging them demanded great skill, and even more so with the first, much prized, wallpapers to reach Europe from China in the 1650s. Not only was each sheet a design in its own right, but the hanger was required to assemble the sheets like jigsaw pieces to form entire landscapes embellished with flowers, birds, butterflies and insects.

Wallpaper needed to combine well with the other elements of a room. 'Although,' says a 19th-century guide, 'it is never, in any tastefully-furnished room its principal feature, it is certainly one of the most important parts of the enrichment, and will invariably be the means of enhancing the whole apartment [room or suite of rooms] or of completely destroying the harmony of the general effect.'

Flock wallpaper, built up in layers to resemble figured velvet and mounted on stretchers, was popular in country houses in the 18th century. The oldest British examples were made in Worcestershire around 1680.

THE SMOKING AND BILLIARD ROOMS ARE CONVENIENTLY PLACED SIDE BY SIDE

Because smoking and billiards were often both enjoyed at the same time, it was logical to have the two rooms adjacent to each other or – in smaller houses – combined into one. Largely male preserves, both could be elaborately decorated.

The fashion for smoking in country houses, and its acceptability, waxed and waned over the centuries. In the late 17th century and throughout the 18th, smoking rooms or parlours, where pipes were enjoyed by men, particularly after dinner, were set aside in houses such as Charborough in Dorset, Kedleston Hall in Derbyshire and Lillingstone Lovell in Buckinghamshire.

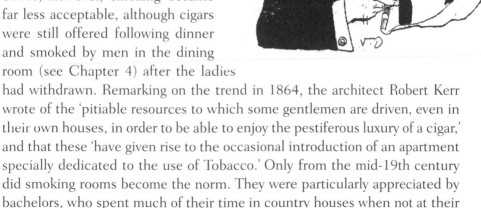

From the beginning of the 1800s, however, smoking became far less acceptable, although cigars were still offered following dinner and smoked by men in the dining room (see Chapter 4) after the ladies had withdrawn. Remarking on the trend in 1864, the architect Robert Kerr wrote of the 'pitiable resources to which some gentlemen are driven, even in their own houses, in order to be able to enjoy the pestiferous luxury of a cigar,' and that these 'have given rise to the occasional introduction of an apartment specially dedicated to the use of Tobacco.' Only from the mid-19th century did smoking rooms become the norm. They were particularly appreciated by bachelors, who spent much of their time in country houses when not at their London clubs.

A popular style of decoration for the Victorian billiard room, which was also used by women (see also Chapter 5) was to have stags' heads plus portraits of horses, or pictures with similar sporting themes, on the walls, around which were raised leather-covered benches for spectators. A high leather stool was provided for the 'nipper' who kept the scores. Walls might also be panelled in wood.

A magnificent summer smoking room in the Moorish style was created over two floors in Cardiff Castle, complete with mythological scenes based on cartoons by the imaginative architect William Burges. The floor tiling depicted a map of the world surrounded by circles of huntsmen, horses, ships and even spouting whales, plus the globes of the medieval universe.

THE TASK OF WARMING A BIG HOUSE IS A LARGE ONE

Indeed it was, especially in the depths of winter, and remained so even with advances in central heating.

Until the 19th century – and often into the 20th – the only means of heating a country house was the open fire. All the rooms, including bedrooms (though not necessarily those occupied by servants), would be fitted with fireplaces, which needed constant attention and daily cleaning if the fires were to stay alight. It was also vital that there should be a draught, which was only too regularly supplied by cold air being sucked in through ill-fitting doors and windows. Equally, smoke often billowed out into the rooms from the downdraughts created in the large, wide chimneys.

In an attempt to improve the efficiency of open fires, the 18th-century physicist and inventor Sir Benjamin Thompson, Count Rumford, discovered that restricting the chimney opening had the effect of increasing the updraught. By inserting bricks into the hearth to angle the side walls and adding a choke to the chimney to increase the speed of air going up the flue, he not only made the fire more efficient but encouraged smoke to go up the chimney instead of billowing into the room. His fireplaces were adapted for use in the kitchen, where they became most popular, but open fires persisted elsewhere.

Hot air 'central heating' systems were also developed, such as the device advertised by one William Day of Lambeth in 1754 'which rarifies cold air until it is hot, and conveys it into Gentlemen's libraries and grand rooms'. At Woburn Abbey such a system was used in the early 19th century, allowing residents and guests the winter-time luxury of strolling from room to room without freezing. Radiators, as we know them today, were also a 19th-century innovation. Among the first to be used in Britain were those installed for the Duke of Wellington at Stratfield Saye in Hampshire.

One of the greatest advantages of the Rumford stove was that it helped to prevent kitchen smells from permeating the rest of the house. In newly built houses of the Victorian era this allowed the kitchen to be sited nearer to the dining room, making it much easier to serve food piping hot.

THE INSTALLATION OF BATHROOMS AND WATER CLOSETS CAN IMPROVE SANITATION SIGNIFICANTLY

Compared with those in cities, country houses were slow to install improved indoor facilities, but houses such as Chatsworth in Derbyshire were at the forefront in the adoption of new technology.

By 1695 Chatsworth had at least ten water closets, made of cedar with alabaster bowls and brass fittings. It was well ahead of its counterparts – the

owners of some country houses were still using earth privies even a century later. When, in the late 17th century, cold baths became fashionable, many houses built bath houses in the grounds as well as having plunge baths indoors. The bath itself might be enclosed in a building of some kind, placed in a grotto or simply out in the open air like a swimming pool.

For bathing, portable hip and sponge baths were used in bedrooms until improvements in plumbing made indoor bathrooms a practical possibility. At Carshalton House in Surrey, completed in 1720, the tiled bathroom (along with an orangery and greenhouse) was placed adjacent to an engine room for heating the water, while at Blenheim in Oxfordshire a bathroom supplied with hot and cold running water was installed directly below the Duchess of Marlborough's bedroom, connected to it by a discreet staircase. The Duke, meanwhile, had his own ornately decorated water closet alongside his dressing room, with marble floor and walls and a gilded ceiling.

With advances in Victorian technology, bathrooms and water closets became much easier to install. New houses such as Bearwood in Berkshire, begun in the 1860s, had no less than 22 water closets and 5 bathrooms, with water pumped from an engine house in the kitchen court to the top of a water tower in the garden. At the same time, great improvements were made in older houses, such as Cardiff Castle, where a Roman marble bath, inlaid with fishes, newts and an octopus made in metal, was converted to use running water.

> *At Osborne House on the Isle of Wight, Queen Victoria and Prince Albert each had their own dressing room, fitted with a bath as well as a shower.*

THE COUNTRY HOUSE IS MUCH IMPROVED WITH MODERN COMFORTS

The advent of electricity and the telephone added greatly to the ease of running a country house, although in its early days electricity was considered somewhat nouveau riche, *if not downright vulgar.*

Cragside in Northumberland was one of the first houses in Britain to be lit by electric light. Here in 1878, aided by the work of the Sunderland-born scientist Joseph Swan (inventor of the first practical light bulb), Sir William Armstrong installed a small hydroelectric plant on his estate to generate electric light in his picture gallery. A year after electricity in Britain was first publicly supplied in 1881, Hatfield House in Hertfordshire had electricity installed, both in the house itself and across the estate. Arriving there after sunset one day in 1890, the author Augustus Hare described the sight: 'All the windows blazed and glittered with light through the dark walls; the Golden Gallery with its hundreds of electric lamps was like a Venetian illumination.'

Despite its impressive technological advances, electricity remained extremely expensive until the early 20th century, making gas the more popular choice for most households. However, with the wider availability of electricity, which coincided with the beginning of the Arts and Crafts movement, from the Edwardian period there was a steady proliferation of new 'electroliers' replacing gas fittings or gasoliers.

For the master of a country house who had business in London or other cities, the arrival of the telephone improved communication immeasurably. As with electricity, Hatfield House was a pioneer of this new technology. At the turn of the 20th century the Duke of Portland, in his home at Welbeck Abbey in Nottinghamshire, even employed his own telegrapher, plus six engineers to care for his electric plant. On a domestic level, *Our Homes and How to Make them Healthy* commended a telephone for communication 'between a gentleman's residence and his stables or entrance lodge'.

Electricity was regarded with suspicion, especially by the less educated country house staff; some of them even thought that it had evil powers.

IN EVERY HOUSEHOLD SOME APARTMENT SHOULD BE REGARDED AS THE SICK-ROOM

When illness struck, patients might also retire to their bedrooms, and would be treated by a sick-nurse. Because so many infections could be fatal, sick-room hygiene was all-important.

The best employers took good care of the medical needs of their household, some even providing a salary and a house in the grounds for a doctor to attend to both family and staff. In the mid-18th century the Duke of Bedford insisted that all his servants were inoculated against smallpox.

'Under all circumstances,' said Mrs Beeton, 'the sick-room should be kept as fresh and sweet as the open air, while the temperature is kept up by artificial heat, taking care that the fire burns clear and gives out no smoke into the room; that the room is perfectly clean, wiped over with a damp cloth every day, if boarded; and swept, after sprinkling with damp tea-leaves, or other aromatic leaves, if carpeted; that all utensils are emptied and cleaned as soon as used, and not once in twenty-four hours as sometimes done ... A careful nurse, when a patient leaves his bed, will open the sheets wide, and throw the clothes back so as thoroughly to air the bed. She will avoid drying or airing anything damp in the sick-room.'

There were other, copious instructions to hand for the positioning and running of the sick-room. Ideally, the room would catch the morning sun, it being 'easier to exclude the sun's rays than to dispense with them'. The bed needed to be placed out of a draught but away from the wall, and all valances and other hangings removed to improve ventilation. Flowers were permitted, particularly after Florence Nightingale declared them to be beneficial because they 'actually absorb carbonic acid and give off oxygen'.

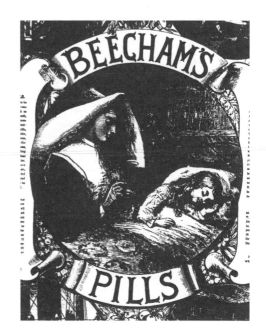

THE IDEAL NURSE

In case of illness a country house family would employ a qualified chamber nurse to tend the sick. She might well have to prepare medicines, weighing out 'ingredients' with great accuracy using apothecaries' scales and liquid measures. The Complete Servant includes the following guidelines for assessing her virtues and behaviour:

- Good temper, patience, watchfulness and sobriety are the cardinal virtues of a good nurse, and when possessed by one who unites skill with those personal qualities, she is a treasure above all price.

- She ought to be past the middle age, and if a married woman or widow, so much the better.

- She ought to be clean in her person, and neat in her dress, and free from habits of drinking or snuff-taking.

- She ought also to be a woman of cheerful and equable temper, and above all things, free from superstition, or belief in charms, omens, signs, dreams, and other follies of gross ignorance.

- Quietness in every respect is of the first consequence … Talking loudly or whispering, so as to excite the suspicion of the patient, should be equally avoided.

- The nurse should scrupulously obey the instructions of the medical advisers, not only as the most likely means of promoting the speedy recovery of the patient, but to remove from herself all responsibility and blame.

Bells are an essential means of communication

They most certainly were from the 1760s, when the bell pull was invented. Properly wired, arrays of bells allowed people in every room to be in contact with servants.

The standard set-up was to have cords or buttons in the upstairs rooms and a row of bells in the servants' quarters, each one numbered to indicate the room in which it had been rung. As systems became more sophisticated, it became possible to connect all the main rooms, including the bedrooms, to the indicator board, and with the arrival of electricity, more modifications could be made. As *Our Homes and How to Keep Them Healthy* of 1883 says, 'Electric and pneumatic bells are free from the inconvenience so common with ordinary bells – namely the stretching of the wires to such an extent as to involve difficulty in moving the bell sufficiently to make it sound.' As to reliability, it says: 'The mechanism of the electric and pneumatic bell, especially the former, is far more simple than that of the ordinary bell, with its innumerable wires, cranks, levers, &c.'

In smaller houses, speaking tubes were installed, using iron or gutta percha pipes into which messages could be shouted – and of course could be heard by all.

Cassell's Household Guide advises: 'Where a number of bells from different parts of the establishment are all brought together, they should be arranged on a bell board in a regular and systematic order – that is, the smallest and highest tones, should be at one end, and gradually range up to the deepest tones.' At Manderston in Berwickshire the lever beside the fireplace in the morning room rang one of 56 bells, each with a subtly different tone, ranked outside the housekeeper's room. The system is still in good working order.

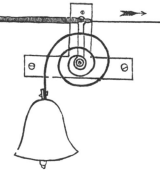

COUNTRY HOUSE GUESTS SHOULD NEVER BE EXPECTED TO SHARE A ROOM

A dictum dating from the 18th century and the rise of the country house party. The exceptions were bachelors, who might be accommodated in dormitories known as 'barracks'.

In the early days of the country house, a bedroom or bedchamber was one of a suite of rooms, usually at ground floor level, and far from private. Here the occupant would not only sleep but dine and receive visitors. Privacy was to be had behind the curtains of the four-poster bed which by Tudor times was, in the wealthy country house, an enormous structure, with massive, richly carved pillars, embroidered and embellished curtains and the family coat of arms displayed at the bedhead.

As country house residents gradually spent more time in communal rooms, so bedrooms shrank, including the main or 'best' bedroom, and were moved to the upper floors. They still had dressing rooms attached, but were no longer public in the way they had once been. For guests, bedrooms for married couples and single women were reasonably comfortable, and furnished with a couple of chairs and a writing table in addition to the basic bedroom furniture.

In Victorian times, and often in later years, a bedroom would also be supplied with a washstand on which a basin, jug and dishes for soap and a

sponge were placed. Ideally, advised *Our Homes and How to Make Them Healthy*, these '... should be of the simplest possible construction, fitted up with tiles to a height of eighteen inches or two feet at the back, with as little woodwork as possible to get wet and dirty ... the centre might be fitted with a zinc receiver to take off the waste water ...'

The accommodation provided for visiting bachelors was much more sparse and might be located in a separate wing or on the top floor of the house, the express purpose being to reduce their accessibility to the female residents and guests. At Kingston Lacy in Dorset, bachelors occupied 'tent rooms' on the attic floor, whose sloping walls were draped with striped awnings.

'One of the silliest pieces of finery in a bed room,' says Mrs Caddy in her Household Organisation *of 1877, 'is the trimmed towel-horse cover. Towels cannot possibly dry if the evaporation is stopped, and even when the cover is made of thin muslin, it is only a troublesome frivolity.'*

∞ Bedrooms for Servants ∞

Servants, who often slept in dormitories, could not be expected to have the same kind of furniture as the gentry, and in fact their rooms were pared to the bare minimum. Cassell's Household Guide, advocating that a servant's room 'should have as few articles in it as consistent with comfort', suggests the following, which to some would have seemed the height of luxury:

A bed and bedstead with two soft mattresses
A pillow, three blankets, two soft under-bleached sheets and pillowslip
Soft and inexpensive coloured counterpane
Chest of drawers
Looking glass
Wash-stand with the usual requisites of white ware
Chair
Eight-day brass clock, locked in a box with a glass cover

… and, it added, 'the less carpet laid on the floor of a servant's room, the healthier and freer from dirt it will be.'

THE LIBRARY SHOULD BE A MOST HABITABLE ROOM

From the mid-17th century, the country house library became a place for relaxation as well as reading. With books often stamped with the family crest, it was also a status symbol.

The earliest country house libraries, like that of Sir William More at Losely in Surrey – which in 1556 contained 273 books – were private rooms adjoining the principal bedchamber. The books, being valuable possessions, were not openly displayed but kept in closets to protect them from both damp and smoke from open fires.

As family book collections grew, libraries became communal rooms, although they were initially relatively small, and as well as serious tomes soon included lighter reading such as novels and plays. In the evenings the library might be used as a family sitting room, as Lady Grey of Wrest in Bedfordshire described in 1745: 'You can't imagine anything more cheerful than that room, nor more comfortable than reading there the rest of the evening.'

By the late 18th century the library had become a place of entertainment, complete with games and scientific toys. Billiard tables were also installed. Writing in 1818 of Bowood in Wiltshire, Maria Edgeworth (Jane Austen's favourite novelist) says: 'The library though magnificent is a most comfortable habitable looking room … it was very agreeable in the delightful library after

breakfast this day – groups around various tables – books and prints – Lady Landsdowne found the battle of Roundway for me in different histories …'

But the library still had a serious purpose, particularly in the Victorian era when the value of education was becoming more appreciated. Advising the aspiring country house owner on the construction of a library, *Our Homes and How to Make Them Healthy* says: 'The principle purpose for which a library is needed being the storage and study of books ... it must be dry and well lighted, but not exposed to direct sunlight ... Plenty of wall-space must be provided for bookcases, and the bookcases should be all arranged for, and form part of, the permanent architecture of the room. The windows, if deeply recessed, and of sufficient width, form convenient places for reading or writing, and, when so arranged, should each have a broad seat.'

The need for comfort in a library is underlined by a Ham House inventory of 1679 which included '2 sleeping chayres, carv'd and guilt frames, covered with gould stuff with gould fringe': still in the house today, they are in fact wing armchairs with reclining backs that can be adjusted with ratchets.

PORTRAITS CAN BOTH COMMEMORATE AND ENTERTAIN

Collecting and displaying portraits in the country house – not only of the family but of other notable figures – began in the 16th century and has continued ever since.

In Elizabethan times, portraits, often bearing the family crest, were hung in the gallery to display the family's significance and mark their wealth. As galleries developed, so full-length portraits began to be painted to embellish them, as at Penshurst Place in Kent, where the Sidney family portraits remain on display. Portraits were also intended to remind viewers of the great deeds of those depicted and to act as an inspiration. Remarking on a visit to Swakeley's in Middlesex in 1665, Samuel Pepys wrote in his *Diary*: 'Pretty to see over the

skreene of the hall … the King's head, and my Lord of Essex on one side, and Fairfax on the other; upon the other side of the Skreene, the parson of the parish, and the lord of the manor and his sisters.'

Country house galleries often contained busts of eminent figures the occupants admired, from Martin Luther and Thomas Cranmer to Charles James Fox and William Pitt.

Like other household objects, portraits in oils needed to be well cared for. Damp was known to be particularly injurious, rotting both canvases and wooden frames. Equally, pictures needed to be hung away from direct sunlight, which faded the colours and made paint blister and crack. Regular dusting was required to keep paintings clean, but little else was advised. *Cassell's Household Guide* warns specifically against the use of soap, which, it says, 'is liable to assimilate with the paint on the picture and to make a lather of the colour itself.'

STAIRS TO SERVANTS' QUARTERS SHOULD NEVER BE OBVIOUS

An architectural dictum from the Victorian era on the discreet placing of the back stairs, in contrast to the house's grand staircases, which, aside from their practical purpose, were symbols of wealth and elegance.

Invisible servants' staircases were a consequence of the 'zoning' of houses, allowing staff access to different floors without being seen by the family or their guests. As the late Victorian designer W.R. Lethaby said, 'It was the affectation of the time that work was done by magic; it was vulgar to recognize its existence or even to see anybody doing it.'

The very earliest Norman country houses had simple, narrow stone staircases spiralling to the upper floors. The grand staircase built around an open well was an innovation of the Elizabethan age, as at Burghley in Lincolnshire where, in imitation of the French style of the time, the wide steps, connected by landings, extended not only to the grand chamber on the first floor but right up to the roof. The trend continued in the Jacobean period, when staircases became ornately decorated, as at Hatfield House in Hertfordshire.

Classic elegance was the hallmark of the Georgian staircase, which was built using cantilever technology and decorated with delicate carvings. More glamorous yet were the staircases of the Regency period, in which a single central staircase branched into two, one branch leading to each side of the upper floor, as at Heaton Hall in Lancashire, designed by James Wyatt.

The staircase was the perfect place for pageantry, notably at the height of the country house seasons of the 19th and early 20th centuries, when guests might well desire to 'make an entrance'.

The exceptions to the ostentatious style of a grand main staircase are in houses of Palladian design, as at Sudbury Hall in Suffolk, in which staircases were placed in corners so as not to spoil the 'harmonic proportions' of the interior.

Nursery furniture should be as plain as possible, easily cleaned and movable

A saying reflecting the attitude towards the country house nursery and the close attention to preserving health in an era when deaths in childhood were commonplace.

In a large house, the children's nursery would consist of two or more upstairs rooms conveniently close to those of the servants, although this arrangement meant that a great deal of time and energy was spent carrying items up and down stairs. However, it was deemed unsanitary for sinks to be situated on the same floor as the nursery, chiefly to avoid the temptation of emptying chamber pots into them. Or, as the 1883 manual *Our Homes and How to Make Them Healthy* said: 'The manifest convenience of having a sink near to rid the nursery department of soiled water has to be weighed against the tendency of all servants to misuse such convenience, and it is best to decide against such sources of mischief.'

The ideal nursery bedroom was spacious and airy, with a high ceiling, and was never used during the daytime except for naps. And, as *The Complete Servant* of 1825 prescribed: 'No servants should sleep in the same room, nor ought any thing to be done there that may contaminate the air, in which so great a portion of infantine life is to be spent. The consequences of vitiated air in bed-rooms,' it warned, are 'often fatal. Feather-beds and bed curtains,' it continued, 'ought to be proscribed, as tending to debility; neither ought the beds to be placed too low, as the most pernicious stratum of air is that nearest the floor.'

Nursery furniture was plain and undecorated, consisting of cupboards and wooden high chairs and cots that could be easily cleaned. The open fire was protected with a fender of wire mesh topped with a brass rail. Windows were guarded with bars, or the lower sashes were nailed down to prevent them from being opened.

The simple 'citizen's furniture' designed by Philip Webb and made by the company founded by William Morris in 1861 was regarded as highly suitable for a Victorian nursery. Also popular was bentwood furniture, such as the rocking chairs introduced in the 1830s by Messrs Thonet of Vienna.

AN ENTIRE SUITE OF ROOMS MAY BE SET ASIDE FOR ROYALTY

A common custom in large houses frequented by royal visitors. Accommodation with sufficient privacy also needed to be secured for their staff.

Detailing the requirements for 'entertaining their Majesties', Mary Spencer Warren, writing in *The Lady's Realm* of 1904, specified that separate suites should be provided for the King, the Queen, and any other members of the Royal Family present and should consist of 'a sleeping apartment, a dressing-room, a bath-room, breakfast-room, study or writing-room and a drawing or reception-room'. These, it advised, should be 'furnished in accordance with the well-known tastes of these distinguished personages' and be 'quite remote from those occupied by other guests in order that perfect quiet and retirement may be secured, as well as facilities for the transaction of government and other state business.'

Royal suites might be exquisitely furnished. Among the items in the state bedroom at Lowther Castle in Cumbria in the early 20th century were

embroidered silk Japanese hangings, furniture originally from Versailles and a carpet woven on the Lowther estate. In houses without royal suites, a huge amount of work was needed ahead of a royal visit to ensure that everything was in perfect order: the preparations might involve the complete redecoration of rooms and re-upholstering of furniture.

Regarding royal staff, it was necessary for private secretaries and the like to be able to come and go easily, not least because the post and telegraph would still bring daily duties for the monarch during the visit. Arrangements for most royal staff, made by the steward or housekeeper, had to comply with the need for both privacy and staff etiquette. As *The Lady's Realm* article points out, 'There is much etiquette of place in the Royal household, certain of the staff taking their meals in certain rooms, and those in the [servants'] hall each taking his or her appointed place at the table in order of precedence and length of service.'

If there was direct access to the garden from the royal suites, then this would be kept exclusively for royal use during the visit.

SPRING CLEANING IS DONE WHEN THE FAMILY IS AWAY FROM HOME

Come the summer, the house would be empty enough for the staff to knuckle down each year to the business of a thorough spring clean.

To retain their warmth, country houses were kept tightly shut during the winter months, during which huge amounts of dust and dirt would accumulate. Even thorough daily cleaning could not rid rooms of the inevitable debris produced over the months by open fires, candles and oil lamps. The beginning of the annual summer Season in London, beginning in May with the Private View at the Royal Academy, was the signal for spring cleaning to start in earnest in the country house.

With the arrival of electricity, gadgets such as the vacuum cleaner became invaluable for the annual spring clean.

Staff engaged in spring cleaning always began at the top of the house and worked downwards. All the carpets were taken up and beaten outdoors in the fresh air. Wooden furniture was French polished and repaired as necessary, and upholstery cleaned – possibly using Fuller's earth, a mineral powder able to absorb grease. Then, unless rooms were going to be used again, all the furniture would be draped with dust sheets.

Grates, lamps and all ornaments were cleaned and polished with extra thoroughness, and decorative fire screens were placed in grates, with flowers added in any rooms that were in use. To complete the work, every window in the house would be cleaned and polished.

CHAPTER 2
THE DAILY ROUTINE

From early morning until late at night country house servants were tirelessly engaged in the myriad tasks needed to keep the home clean, tidy and running smoothly, and the precise timetable for each hour of the day – and each day of the week – varied little from house to house.

The details of the routine varied, however, for each member of the household staff, for while maids were busiest in the mornings, preparing for the business of the day, the services of footmen were required early in the day and throughout the afternoon and evening. For valets and ladies' maids the day could not end until the master and mistress of the house had retired to bed, while the butler needed to make sure that the house was properly secured before he could finally get some rest. Laundry was seen to on several days of each week, beginning on a Monday, and time was set aside, particularly during the summer months, for activities such as making preserves and salting meat to ensure supplies throughout the year. During the shooting and hunting seasons, and at Christmas and times of family celebration, there were many special meals to prepare.

Whatever the time of day, the sound of a bell from the room-by-room array in the servants' quarters demanded immediate attention. For the family, the day also ran to a strict routine, with set times for meals, prayers and leisure activities. Callers would always be received, and calls to neighbouring houses made, in the afternoon. When there were children in the house their day was rigidly organized by the governess with set times for meals, lessons and play.

THE SCULLERY MAIDS MUST BE THE EARLIEST RISERS

Inevitably, as it was the duty of these lowliest servants to get the country house ready for the day, beginning at about 5.30am.

Once she had risen from her small, sparsely furnished quarters under the eaves of the house, the first early morning task for the scullery maid, who might be only 13 or 14 years old, was to scrub and sweep the kitchen areas and to clear the fire grate of ashes if this had not been done the night before. She would then wipe over the kitchen range, which also needed to be blackleaded every two or three days. The fire was then re-laid and lit in the range and fires also lit under any coppers or boilers for heating water.

The scullery maid's next duty was to heat and distribute hot water for the senior servants and to lay breakfast in the servants' hall before she began her daily chores of washing up plates and dishes, and, as detailed in *The Complete Servant* of 1675, taking care that all utensils were '*always kept clean, dry* and *fit for use*'. She was also expected, the manual continues, 'to assist the kitchen maid in picking, trimming, washing and boiling the vegetables, cleaning the kitchen and offices, the servants' hall, housekeeper's room, and steward's room, and to clean the steps of the front door and area.'

While acknowledging that the role of scullery maid 'is not, of course, one of high rank', Mrs Beeton insisted that 'if she be fortunate enough to have over her a good kitchen-maid and clever cook, she may very soon learn to perform various little duties connected with cooking operations, which may be of considerable service in fitting her for a more responsible place.'

FOOTMEN ARE EXPECTED TO RISE EARLY, BEFORE THE FAMILY ARE STIRRING

This was necessary so that they could get the dirtiest jobs completed before being called on to conduct their many other duties.

Expanding on the early morning duties of the footman, *The Complete Servant* lists his jobs as 'cleaning the shoes and boots, knives and forks, brushing and cleaning clothes, hats and gloves and cleaning the furniture, &c. &c.' He might also clean lamps and pump and carry water. To save his livery from being dirtied he was advised to wear 'a pair of overalls, a waistcoat and fustian [heavy cotton] jacket, and a leather apron'.

Contemporary books are packed with instructions for footmen. Once boots and shoes had been attended to they would see to the furniture, using an oil or other preparation of the appropriate colour to treat the different types of wood composing everything from tables and chairs to sideboards. Oil, if it was used, needed to be rubbed off quickly, then each item polished with a clean cloth. Wax was best applied sparingly, and also rubbed off with a separate cloth.

While doing his morning cleaning the footman needed a white apron to hand to put on quickly should he be called away from his duties for any reason, such as to answer the door or wait at breakfast.

Mirrors were also the province of the footman and, being expensive items, needed treating with particular care. The frames would never be allowed to get wet or damp. Early 19th-century instructions run: 'First, take a clean, soft sponge, just squeezed out of water, and then dipped in spirits of wine; rub the glass over with this, and then polish it off with fine powder blue, or whiting tied up in muslin, quickly laid on, and then well rubbed off, with a clean cloth, and afterwards with a silk handkerchief.'

The footman was expected to get through all his dirty jobs before the family began to stir. Then he needed to change into clean clothes in time to lay the table for the family's breakfast. In households that did not employ a valet, the footman's early morning tasks would also include brushing his master's coat and hat, and laying out the clothes he was to wear.

FAMILY PRAYERS ARE SAID A QUARTER OF AN HOUR BEFORE BREAKFAST

This daily gathering, announced by a gong rung by the footman, was attended by all members of the household. Everyone would also attend church on a Sunday.

Prayers would be held in the chapel, if there was one, or in the library, hall or dining room, and would be attended by children brought down from the nursery. Usually held at 8.30 and lasting some 15 minutes, they were led by

the master of the house or, if he was not in residence, by the mistress, with everyone kneeling or standing with heads bowed. Before putting on clean aprons, straightening their caps and going upstairs for prayers, kitchen staff – who would have already eaten – would make sure that everything was ready for serving breakfast immediately afterwards, with cooked dishes kept warm. Following prayers, family announcements – and possibly open reprimands for staff – were made before the servants filed out.

⌁ GOD-GIVEN DUTIES ⌁

The link between service and godliness was avidly encouraged, with displays of religious texts in the servants' quarters, the giving of Bibles as gifts, and even in hymns. The 19th-century churchman John Keble, for instance, wrote the lines:

The trivial round, the common task,
Will furnish all we need to ask,
Room to deny ourselves, a road
To bring us daily nearer God.

Two centuries earlier the Jacobean poet George Herbert had expressed a similar sentiment, in lines also set as a hymn in the 19th century:

Teach me, my God and King,
In all things Thee to see,
And what I do in anything,
To do it as for Thee.

A servant with this clause
Makes drudgery divine;
Who sweeps a room as for Thy laws,
Makes that and the action fine.

For Sunday morning services, whether in the chapel or at a nearby church reached on foot, servants would dress soberly. As an Edwardian servant Doris Bodger recalled: 'We had to wear black coats and skirts, black shoes, stockings and gloves, and a hat which was called a toque which made a young girl of thirteen look like a grandmother.' While the family would sit in pews reserved for them at the front of the church, servants would occupy the back pews or the gallery, seated according to their rank in the household. On alternate Sundays servants might also be expected to attend church in the afternoon or evening, a ritual that greatly reduced their leisure time.

THE BEDROOMS ARE CLEANED AFTER BREAKFAST

A daily duty of the housemaids, which also included attention to the dressing rooms of all other residents in the house. Ladies' maids would perform similar duties for their mistresses.

Seeing to the bedrooms began with opening the windows (or just the door in cold weather) to let the room air, and emptying any chamber pots. The bed covers were stripped off and hung over the backs of chairs to air while the mattress was shaken. The housemaid then cleaned out the fireplace and re-laid the fire before going downstairs to wash her hands and put on a clean apron. On her return she would make each bed before brushing the carpet, dusting the furniture, dusting or washing any ornaments and filling carafes with fresh water. A junior housemaid was required to see to the upper servants' beds and sweep and dust their rooms before cleaning other rooms in the house.

Later in the day, while the family and guests were dining, the housemaid would return to their

After the cleaning had been done the housekeeper inspected each room closely, making sure that towels, soap, writing-paper and inkstands had been attended to and that drawers and wardrobes had been properly dusted.

bedrooms, taking hot water to each as necessary. She would, especially in winter, also add a warming pan to the bed to prepare it for the night. To make sure that a bed had been properly aired she might use this method recommended by *Enquire Within*: 'Introduce a drinking glass between the sheets for a minute or two, just when the warming-pan is taken out; if the bed be dry, there will only be a slight cloudy appearance on the glass, but if not, the damp of the bed will collect in and on the glass…'

∽ From the housemaid's cupboard ∽

In her cupboard, which she was required to keep permanently in good order, the housemaid kept all her cleaning materials. From the 17th century, guides offered all kinds of advice to housemaids, such as:

To dust carpets and floors – *Carpets should not be swept with a whisk-brush more than once a week; at other times sprinkle tea-leaves on them, and sweep carefully with a hair-broom, after which they should be gently brushed on the knees with a clothes'-brush.*

To clean marble – *Mix up a quantity of the strongest soap-lees with quicklime to the consistence of milk, and lay it on the marble for twenty-four hours; clean it afterwards with soap and water.*

To clean floorcloth [*a waxed canvas used in servants' quarters*] – *Wash in the usual manner with a damp flannel, wet it all over with milk and rub it well with a dry cloth, when a most beautiful polish will be brought out.*

If drawers come out stiffly – *Rub over the spot where they press with a little soap.*

STAFF BREAKFAST IS PREPARED BY THE KITCHEN MAID, FAMILY BREAKFAST BY THE COOK

As a rule, staff would breakfast long before the family and guests were served at 9.00am, following family prayers.

At about 6.30 in the morning, the kitchen maid, also known as the under cook, lit the kitchen fire and put water on to boil ahead of preparing tea and toast for the housekeeper and the ladies' maids: this would be delivered to them in their rooms by the housemaids. The kitchen maid then attended to the preparation of breakfast for the rest of the staff, ready for serving in the servants' hall at 8.00am.

In the early morning both kitchen maid and cook were occupied making bread rolls, muffins, oatcakes and crumpets, plus Sally Lunns, bannocks, wholemeal cakes and scones – all to be served warm or super-fresh from the oven or griddle at breakfast upstairs. Toast also needed to be made. Eggs in all forms, including omelettes, were popular for simple breakfasts, and they would also appear as part of expansive meat-laden buffets (see Chapter 3).

To serve breakfast, the butler carried in the tea urn, a footman following behind with a tray of hot dishes. The footman then rang the breakfast gong. It was not customary for everyone to take breakfast at the same time and, says *The Servants' Practical Guide*: 'The master and mistress of a house wait breakfast for neither guests nor for the members of the family, but take their places at once at table. The butler will then ask each person whether they would like tea or coffee, determine what they would like to eat and serve them from a sideboard.'

As a country house guest it was essential to fit in

with the timetable of the house. Given the varied pressures of organizing a country house weekend, Edwardian etiquette books offered sound advice such as: 'If the hostess offers you breakfast in bed, and indeed urges you to have it, accept.'

Additional duties for the kitchen maid were to keep all kitchen utensils clean and to help the cook in preparing meals throughout the day. At all times she worked under the personal – and close – direction of the cook.

DAILY STORES ARE APPORTIONED BY THE HOUSEKEEPER

As the keeper of the storeroom key, the housekeeper was in charge of all provisions, including flour, rice, sugar, tea and coffee.

Stores for each day were given out by the housekeeper once all the arrangements for family breakfast were in order, after which she would check the stillroom (see Chapter 3). Throughout the day she supervised the work of the various maids. In the late afternoon, following tea, she was responsible for pounding and grinding spices, washing and picking over currants, stoning raisins and breaking up the large loaves in which sugar was supplied, then pounding and rolling it into fine grains. She might also pare oranges and lemons and set the rinds aside to dry – they would later be grated and stored for future use as flavourings.

Like the butler, the housekeeper needed to be a person of integrity. As *The Servants Book of Knowledge* said, it was most prudent to choose 'a woman of age and experience who [had] either kept house herself, or been

As well as provisions, the housekeeper was responsible for looking after all the linen of the house, of which she would keep a detailed inventory.

long in the service of others.' A mature woman, both in age and demeanour, was essential, and as a sign of her status the housekeeper was always addressed as 'Mrs', whether or not she was married.

ODE TO A HOUSEKEEPER

The perfect housekeeper was lauded in verse in the 1850s by Philip Yorke II of Erdigg in North Wales. A competent cook might well be promoted to this role.

Upon the portly frame we look
Of one who was our former Cook.
No better keeper of our Store,
Did ever enter at our door.
She knew and pandered to our taste,
Allowed no want and yet no waste.

KNIVES AND OTHER CUTLERY ARE ATTENDED TO IN THE BUTLER'S PANTRY

Cleaning and sharpening knives was a morning task for the hallboy or a footman. Knife cleaning might be also carried out by the under gardener.

Seeing to cutlery was a lengthy business in any large country house. Knives, which stained and rusted easily before the advent of stainless steel, needed particular attention. First they needed to be carefully washed in hot water with a little salt added to remove every trace of grease, but without immersing their handles, which, if made of ivory, would become discoloured and loose if placed in hot water. The blades were then thoroughly dried before being rubbed on a knife board covered with brick dust or, in later times, pushed into a rotary knife cleaning machine into which a patent powder had been shaken (see box). Any knives that were not going to be used again immediately would have their blades rubbed with mutton suet to exclude air and prevent them from rusting.

Silver cutlery was the footman's responsibility. It was washed in hot water every day and wiped, then buffed up with a soft rag or leather. Weekly – or more often, depending on how much it was used – it was cleaned with a paste made from whiting mixed with ammonia and cold water or with rags boiled in a mixture of hartshorn powder (ammonium carbonate) and water.

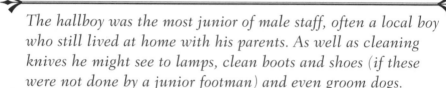

The hallboy was the most junior of male staff, often a local boy who still lived at home with his parents. As well as cleaning knives he might see to lamps, clean boots and shoes (if these were not done by a junior footman) and even groom dogs.

∞ HELP FROM GADGETS ∞

Machines to assist with cleaning knives and other cutlery were as great a boon as this 1850s advertisement from The Illustrated Times *suggests:*

KNIVES CLEANED *by the old brick-dust board are rapidly worn out, by some machines they are notched or blunted, by others no time or labour is saved, while all are very expensive; but by the Patent Press Knife and Fork Cleaner, which cannot get out of order, all these evils are avoided, at a price to suit the most economical. Prices: To clean six knives, &c., 14s; 12 ditto, 20s; 18 ditto, 25s; 24 ditto, 30s. Oliver Long, patentee and manufacturer.*

Such machines needed to be used with care – if pushed in too far the knives would get worn at the shoulder and their edges would be blunted.

WINE FOR THE DAY SHOULD BE DECANTED IN THE MORNING

Just one of the many duties of the butler, who was responsible for everything to do with alcoholic beverages. Decanting was essential to separate off deposits or dregs accumulated within bottles.

The country house cellar was the butler's domain, and his cellar book the record of every bottle held within it (including sherry and port as well as wine and champagne) and its date of entry and exit. Wine purchase, however, was the responsibility of the master of the house, who was, unless he trusted his butler implicitly, keeper of the key to the wine cellar. 'It is the butler's duty,' says *The Servants' Practical Guide*, 'to decant wine for daily uses, both for luncheon and dinner, and to put away the decanters of wine after each meal. On the occasion of a dinner party or dance,' it adds, 'a master of a house gives out so many dozens of wine, according to the number of guests invited.'

> *When decanting, the butler was advised to use a wine strainer lined with a fine material such as cambric. The art lay in pouring the wine into the decanter slowly and steadily.*

Because of his position, the butler needed to be a man of integrity, but insobriety was a common failing of butlers, and one for which dismissal was inevitable if discovered. He also needed to be sure that drinks never fell into the wrong hands. As Anthony Heasel says in his *Servants Book of Knowledge* of 1773: 'Take great care of your wine and other liquors ... not only to keep them in good order, but likewise to prevent their being embezzled, or given away to any person besides those who have a right to them according to your instructions.'

Correct Treatments for Perfect Wine

To clarify and stabilize it, wine was 'fined' by the butler on a regular basis, then bottled by him, the latter often being done as early as four or five in the morning to avoid its interfering with other cellar work. These instructions are based on those recommended by Mrs Beeton:

Fining

1. Draw off a gallon of wine from the cask and whisk a quart of it with the whites of four eggs.
2. Pour back into the cask through the bunghole.
3. Stir up the whole cask, using a clean split stick moved in a 'rotatory directon'.
4. Pour in the remainder of the wine drawn off.
5. Stir again, skimming off any bubbles that rise to the surface.
6. Close the bunghole and leave to stand for three or four days.

Bottling

Best done by two people, one to draw off the wine, the other to cork the bottles.

1. Thoroughly wash and dry the bottles.
2. Soak corks in hot water.
3. Bore a hole in the lower part of the cask with a gimlet, 'receiving the liquid stream which follows in the bottle and filterer, which is placed in a tub or basin' to avoid waste. As the bottom of the cask is reached, put a thick piece of muslin over the filter to prevent 'viscous grounds' from entering the bottle.
4. Make sure each bottle is filled 'up to the mark' but not too full.
5. Drive in the corks with a wooden mallet.
6. Count, record and store the bottles.

ATTENTION TO LAUNDRY BEGINS ON A MONDAY MORNING

It is no myth that before the advent of the washing machine laundry took most of the week to complete. Monday was the usual day for it to start.

Big country houses had laundries detached from the main building. The largest consisted of a series of rooms including a wash house or washing room, drying room and hot air closet, mangling room and an ironing and folding room. Several laundry maids would begin their duties by collecting all the laundry and sorting it by fabric – white linen and collars; sheets and body linen; fine muslins; coloured cotton and linen; woollens; and coarse kitchen and greasy cloths.

Soaking was the next task, for all but the finest items and the woollens. This was done by folding the articles, then immersing them in lukewarm water (ideally rainwater, which was often collected for the purpose) with lye added in washing troughs or buckets fitted with draining spigots at the base. When buckets were used the liquid was drained off through the spigot and the soaking repeated until the liquid came out clean. Except for the dirtiest clothes they could then just be rinsed and dried before being starched, if necessary, and ironed.

Keeping track of the laundry was ultimately the responsibility of the housekeeper, but laundry maids needed to keep exact lists of everything sent to the laundry in the 'washing book'.

Lye was made by putting wood ash, collected from ovens, furnaces and fireplaces, into a sieve and pouring water through it. The liquid was then strained through muslin. For the best, whitest wash, apple wood was prized, but oak lye was the strongest of all. For stains, very greasy items were rubbed with soap before being washed, while ink stains were treated by being rubbed with oxalic acid then rinsed. For fruit and wine stains Mrs Beeton recommended dipping items in a solution of sal ammonia or spirits of wine before rinsing.

⌘ THE SCALE OF THE WASH ⌘

The inventory of linen for Shugborough in Staffordshire gives an idea of the scale of laundry tasks ('diaper' and 'bird's eye' are patterned weaves):

Damask tablecloths	85	Damask table napkins	92 dozen
Damask breakfast cloths	23	Damask tea napkins	29 dozen
Diaper tablecloths	67	Diaper table napkins	44 dozen
Bird's-eye tablecloths	87	Bird's-eye table napkins	21 dozen
Huckaback towels	32	Diaper towels	5
Holland sheets	17 pairs	Holland pillowcases	17 pairs
Second sheets	19 pairs	Second pillowcases	18 pairs
Servants' sheets	24 pairs	Chintz quilts	2
White quilts	3		

TUESDAY AND WEDNESDAY ARE THE DAYS FOR WASHING AND DRYING LAUNDRY

After Monday's soaking came more arduous tasks for the laundry maids.

Firing up the copper boilers was the prelude to washing heavily soiled items, which were immersed and rubbed clean. Soda might be added to help the process, or items could be rubbed with hard yellow soap. Laundry maids suffered greatly from constantly immersing their hands and arms in the hot water, but were helped by the use of washboards and dollies – long-handled wooden objects with four or five legs that looked rather like upside-down milking stools. The laundry was then rinsed in cold water and, to remove excess water, items were either twisted – often with one end attached to a hook on the wall – or, even better, passed through the rollers of a hand-cranked wringer.

Because soap was expensive it was stored in a dry place until it reached maximum hardness. To make a soap jelly, perfect for the finest, most delicate fabrics, shavings were taken from the corners and edges of the blocks and dissolved in water.

For drying, hanging laundry such as sheets and tablecloths outside where sunlight could bleach out any stains was ideal. Giving advice for laundry maids, *Cassell's Household Guide* says that: 'Sheets and table-cloths should be hung with the short side towards the wind, to enable the air to blow the folds apart. Shirts should be suspended from the bottom hem.' Anything unsuitable for hanging on a line with wooden pegs was put on racks or clothes horses in the drying room.

In inclement weather, all the laundry had to be dried indoors. From the mid-19th century country houses had drying closets – brick enclosures under which ran the hot water pipes or the furnaces used for heating the house. Inside were long iron frames that slid in and out. The air needed to circulate freely within the closets to prevent laundry from turning yellow.

MANGLING, STARCHING AND IRONING ARE CARRIED OUT AT THE END OF THE WEEK

These were the last stages of the laundry process, usually done on Thursdays and Fridays, and performed well away from the wash house.

The mangle, though now associated with removing water from clothes, was an invention used for flattening washed and dried sheets and tablecloths, and any simple clothes without frills. In the box mangle, clothes were wrapped around wooden cylinders and rolled to and fro by turning a wheel, with pressure applied from above. Alternatively they could be fed through the rollers of an upright mangle.

After mangling, linens would be flattened further in a linen press.

Starch was a vital laundry product, used for stiffening everything from linen tablecloths to shirt fronts and collars before they were mangled, pressed or ironed. Glenfield Patent Starch, used in the Royal Laundry of Queen Victoria, was widely advertised as available from 'all Chandlers, Grocers, &c., &c.', but some households made their own starch. For shirt fronts, a cold starch made by melting wax in water was used, while for larger items a paste of ready-bought starch and hot water

For ruffles and flounces a gauffering iron was needed. This was a gadget like a metal test tube set horizontally on a stand, heated by inserting a metal poker-like rod fresh from stove or hearth, around which frilled cuffs and collars could be curled. For rows of frilled trimmings, heated gauffering or crimping tongs were used.

was preferred. For economy, laundry starch could be made from ground rice, wheat or potato flour.

Any necessary ironing was carried out with flat irons heated on a stove or with box irons, which had cavities into which slugs of metal, heated on the fire, were inserted. Alternatively, there might be a receptacle at the back of the iron designed to contain heated charcoal.

YOUNG CHILDREN MUST BE TAKEN OUT FOR A WALK EACH MORNING AND AFTERNOON

Part of the unbreakable daily routine of the nursery, which, as in the rest of the house, began in the early hours.

The most junior member of the nursery staff, the nursery maid, would be the first to rise, at about 6.00am, when she would light the fire and complete any cleaning before the children rose at around 7.00. By 7.30 children would be bathed and dressed before breakfast, prepared by the nursery maid, was served.

A walk, taken at 9.30 in summer and 11.00 in winter, was the main activity of the morning but not, advised *The Complete Servant*, 'long enough to fatigue them'. If there were two or more children then the nursery maid would accompany the nurse. On their return, older children would attend to their lessons while younger ones lay down to rest.

Following their main meal – originally known as 'dinner' – and weather permitting, children were taken out again for an hour and a half in the afternoon, then washed and tidied up before being taken by the nurse to the drawing room at 4.30. There they would be left with their mother (and possibly their father) for about half an hour before returning to the nursery for their

 THE NURSERY MAID'S DUTIES

The Servants' Practical Guide *lists many other duties for the nursery maid, these being 'of a very practical and subordinate character', including:*

- *Sweep and dust the day nursery, clean the grate and light the fire.*

- *Light the fire in the night nursery when one is required.*

- *Bring up the water for the children's baths. Assist the nurse in washing and dressing the children*

- *Lay the nursery breakfast table and afterwards clear away and wash up the breakfast things.*

- *Make the beds and empty the baths.*

- *Sweep and clean the night nursery.*

- *Set the nursery dinner-table, and bring up and clear away the nursery dinner, including that of the nurse and her own.*

- *Play with and amuse the children before tea.*

- *Assist the nurse in preparing the children's things for wear the next morning.*

tea. 'Sometimes,' said *The Servants' Practical Guide*, 'children are brought down after instead of before tea, five o'clock being the hour for nursery-tea; but as half-past five is near to the hour of young children's bed-time, they are rather inclined to be cross and sleepy if taken to the drawing-room after they have had their tea.'

After the children had been bathed and put to bed at 6.00pm, the nurse was brought her supper by the nursery-maid or under housemaid. She would then spend her evening mending and repairing children's clothing and cutting out and making new garments.

Nurses were highly regarded members of the country house staff, usually well treated by their mistresses. This, it was said, tended to make nurses 'tyrannical and overbearing'.

A GOOD GOVERNESS ADHERES TO A STRICT TIMETABLE

As in the nursery, the schoolroom timetable, set to synchronize with the routine in the rest of the house, was strictly kept.

Even before breakfast, older children might be obliged to practise the piano or another musical instrument. With breaks for meals and walks, lessons could last for up to eight hours a day, continuing even after nursery tea and time spent with parents.

Expanding on the qualities of the perfect governess, *The Complete Servant* said that: 'In addition to a thorough knowledge of the English Language, and to the power of being able to write a letter in a graceful and accurate style, the governess ought to be moderately acquainted with the French Language; and it would be an advantage if she knew something of Italian, as the language of music. She ought also be able to play the pianoforte...

It will be also expected that she shall be able to teach the elements of dancing, at least, the steps and ordinary figures of fashionable practice.'

Nor were more academic studies ignored. The governess was required to be knowledgeable in '… the useful art of arithmetic, the constant exercise of which will so much improve the reasoning power of her pupils,' and there was no reason why she should 'omit to introduce to her pupils the geographical copy books, and other elementary books of geography, by Goldsmith; and the familiar keys to popular sciences'. Other subjects added to the list were drawing and needlework plus 'religion, morals and temper'.

Since many governesses were women who had fallen on hard times, some undoubtedly took out their disappointment and bitterness on the children in their care. Even the most well meaning often found it hard to build a friendship and rapport with their charges, as discipline was paramount, with bad behaviour often severely punished with slapping and beating. Despite this, many governesses were much loved by their pupils.

Even a governess could break the rules. In 1907, when the Kaiser visited Kingston Lacy in Dorset, Governess Tidmarsh added herself to the commemorative family photograph, leading to her dismissal when her 'sin' was revealed.

IT IS CUSTOMARY FOR THE SERVANTS TO DINE WHILE THE FAMILY TAKE LUNCHEON

The main meal of the day for the staff was served in the servants' hall at 1.00pm, even after dinner moved to a later hour for master and mistress.

Usually, both upper and lower servants dined together in the servants' hall, but where there was a house steward, he and the housekeeper might dine together in either of their rooms. Or senior staff might partake of selected dishes only. Commenting on staff meals on a visit to England in the late 18th century, the German novelist Sophie von la Roche said: 'England knows nothing of separate cooking for servants, who partake of all the courses sampled by the masters, the latter having first choice and the servants what remains….' This was certainly true of the senior servants, though not necessarily of the lower ones, who might even be denied proper rations by their superiors below stairs.

Meat was a staple of servants' dinners. At Cannons, the Middlesex home of the Duke of Chandos, each 18th-century servant was served 21oz (600g) of beef on Tuesday, Thursday and Sunday, the same amount of mutton on Monday and Friday; and 14oz (400g) of pork on Wednesday and Saturday. In addition they were given 'pastries, jellies, tarts, and all good things prized by the master'.

Ale and beer were habitually drunk, a pint being the usual serving. In some houses staff were offered 'beer money' in place of the beverage, which might be spent by men in the local hostelry.

ON HER VISITS, THE LADY OF THE HOUSE IS ACCOMPANIED BY A FOOTMAN

A maxim underlining the origin of the footman as a servant who walked or ran alongside a coach. Afternoon was the usual time for ladies to make visits, but the footman might be needed at any time.

In a house with three or more footmen, the 'lady's footman' was the second most senior, and he was spared the dirtiest jobs. When accompanying his mistress in a carriage, a footman dressed in his best livery, with clean shoes

> *A well-prepared lady's footman had umbrellas, rugs – and even a stone hot water bottle – always to hand to ensure his mistress's comfort while travelling. He also ensured that the interior of the carriage was spotlessly clean.*

and well-brushed hat and greatcoat. He received instructions from his mistress as she entered the carriage – climbing the steps he had already put up for her – and delivered these to the coachman. When shutting the carriage door he needed to take extreme care not to 'injure any one, or the dresses of the ladies'.

The lady's footman was also responsible for carrying all her messages and delivering her invitations. Should she wish to go out on foot he would accompany her. He might also prepare her breakfast and wait behind her chair at both breakfast and dinner. If she went out riding he would clasp his hands together so that she could put her foot in them to mount her horse.

AFTERNOON TEA IS SERVED IN THE DRAWING ROOM

Tea was not merely a form of refreshment but an occasion for ladies, especially, to exchange news and gossip and to plan their social lives.

By the Victorian era, when luncheon was eaten in the middle of the day and dinner served in the evening, tea was taken in the afternoon. Mrs Beeton gives a vivid description of the event at which both tea and coffee were served, poured by the hostess: 'As soon as the drawing-room bell rings for tea, the footman enters with the tray, which has previously been prepared; hands the tray round to the company, with cream and sugar, the tea and coffee being generally poured out, while another attendant hands cakes, toast, or biscuits.'

The fashion for taking afternoon tea is said to have been started by Queen Victoria's friend Anna Russell, Duchess of Bedford, who became peckish in the long gap between lunch and dinner.

Tea in the warmth of the house was also relished by sportsmen and women. As Agnes Jekyll says in her *Kitchen Essays* of 1922: 'Hungry hunters and shooters, triumphant from the chase, love to quench their thirst and spoil their dinners under the stuffed heads in the great hall ...' She goes on to recommend a caraway tea bread, a variation on the caraway cake so popular in the 19th century: 'Ingredients: Three teacups of flour, 2 teaspoonfuls baking powder, 1 teacup caster sugar, 1 large dessertspoonful ground caraway seeds,

1 egg, 3oz butter, 1 teacupful of boiling milk. Mix flour, baking powder, and sugar, rub in butter, mix the milk warmed with the egg beaten and the ground caraway seeds. Knead into a flattish brick-shaped loaf or cake, and bake 20 minutes in a quick oven. To be eaten fresh, with a little butter.'

IT IS THE FOOTMAN'S DUTY TO LAY THE DINNER TABLE

Although he would be closely supervised by the butler in doing so. He was also obliged to serve at meals throughout the day.

Before the cutlery, glasses and ornaments were set out, the dinner table was covered with a thick baize cloth over which was placed a fine white damask tablecloth. The footman needed to make sure that this cloth, and the damask napkins, had been properly ironed and aired in the laundry. For a dinner party the usual cover for each person comprised, said *The Servants' Practical Guide* of 1880: 'Two large dinner-knives and a small silver fish-knife; two large dinner-forks and a small silver fish-fork; these are placed on the right and left-hand side of the space to be occupied by the plate, a table spoon for soup is also placed on the right-hand side, bowl upwards.' Dessert spoons and knives for fruit and cheese were handed only when these dishes were served.

Before dinner the footman needed to make sure that he was properly dressed in his livery, with white tie and spotless white cotton gloves.

'A glass for sherry, a glass for hock or claret (whichever is given) and a glass for champagne placed at the right-hand side' are recommended by the *Guide*. 'A tumbler,' it adds, 'is not used at a dinner-party unless a guest does not drink wine, when a tumbler would be asked for of a servant in attendance.' Small engraved carafes, or water bottles, were placed on each side of the table, one to each couple. A silver or fine glass salt cellar was similarly positioned. When all was done the butler would examine the table, measuring each setting with a ruler to ensure that all were perfectly equal. He also placed any precious silver items on the table himself.

⌒ THE DUTIES OF DINNER ⌒

The footman's role in dinner service:

Ring the bell about half an hour before dinner.

Carry up everything that is likely to be needed before and during the meal on trays.

Prepare the drawing room, lighting the fire.

Announce in the drawing room that dinner is served.

Stand behind his master or mistress.

Hand dishes around for all courses.

Take away dish covers removed by the butler.

And following dinner …

Clear the table.

Put away the plate.

Wash the glass and silver.

Prepare and assist in carrying tea and coffee to the drawing room.

Attend to the requirements of the gentlemen in the smoking room.

Be in attendance in the front hall when guests are leaving.

Bring candles or lamps for the members of the household before they retire to bed.

While waiting during dinner the footman was required to be 'obtrusive to none', to hand food to the left side of each diner and to hold dishes so that food could be taken with ease. 'In lifting dishes from the table,' says Mrs Beeton, 'he should use both hands, and remove them with care.' Additionally, he would set and wait at the table at breakfast, and clear up the room afterwards, sweeping up any crumbs, shaking out the baize cloth and replacing the tablecloth. At tea time he carried up the tray holding toast and muffins. He also answered the door, greeting and announcing guests.

A MASTER'S APPEARANCE IS THE RESPONSIBILITY OF HIS VALET

Both at home and away, attending to the wardrobe was a major role for the valet who, in earlier centuries, would also shave his master and attend to all his personal needs.

Early each morning, before his master arose, the valet needed to be sure that everything was in order, with dressing gown and slippers airing before the fire and clean linen perfectly ready. Clothes, brushed the previous night, would be placed over the backs of chairs, right sides out, and boots cleaned by him, not one of the other servants. Clean water for washing would be ready, along with pristine hair, nail and toothbrushes. If a gentleman did not shave himself, then it was essential that the razor was stropped on a leather to sharpen it, having first been warmed by being dipped in hot water. The valet then helped his master on with his clothes.

Once his master had departed for his morning activities the valet busied himself folding away night clothes, clearing and cleaning the dressing stand, replacing used towels with clean ones and dusting the room. He would put out any clothes that might be needed 'in case of a master's coming home wet from a ride'. Clothes for the evening including white tie for dinner were prepared and laid out.

Discretion was a prime quality in a valet. It was advised that he should have 'as short a memory as possible' and 'be very cautious of mischief-making or tale-bearing, to the prejudice of other persons'. The valet needed to be a man of polish, especially when accompanying his master on visits away from the house. On such occasions he packed for his master, including enough clothes for the visit plus special dress for activities such as hunting and shooting. At the destination he then put everything in order, as at home. Out shooting he was required to be on hand to load his master's gun.

> 'Young men who pay rounds of visits to country houses,' says The Servants' Practical Guide, 'cannot easily dispense with a valet,' adding that: 'Sportsmen, and men given to hunting and shooting, find the services of one invaluable.'

THE LADY'S MAID MUST ATTEND TO EVERYTHING REGARDING THE TOILETTE OF HER MISTRESS

All the most intimate forms of attention, to both her mistress's person and her clothes, were the responsibility of the lady's maid.

Several times a day the lady's maid was required to dress, undress and re-dress her mistress, as well as attending to her hair, but her first tasks each morning were to lay out her clothes (if this had not been done the night before), to make sure that she had hot water for washing and to check that the housemaid had laid the fire and attended to her dressing room. Once her mistress was dressed, and her hair combed and styled, the lady's maid needed to put out any clothes that would be required for walking or riding and prepare garments for dinner in the evening.

In helping her mistress to dress, every detail needed attention. For instance, as Mrs Beeton says: 'Arrange the folds of the dress over the crinoline petticoats … See that the sleeves fall well over the arms. If [the outfit] is finished with a jacket, or other upper dress, see that it fits smoothly under the arms, pull out the flounces and spread out the petticoat at the bottom with the hands, so that it falls in graceful folds. In arranging the petticoat itself, a careful lady's maid will see that this is firmly fastened around the waist.'

When a lady kept a dog it was the lady's maid's duty to attend to it, wash it and take it out for walks as necessary.

Cleaning and mending clothes, and attention to millinery, were daytime chores, as were washing undergarments and cleaning shoes. Although she might have free time in the evening, the lady's maid had to stay up until her mistress returned from dinner and any evening entertainment, when she would assist in undressing her, helping her into her nightclothes and ascertaining her needs for the following day. Finally, all her jewellery was put safely away.

THE MISTRESS'S WARDROBE IS KEPT IN GOOD REPAIR BY HER MAID

Caring for her mistress's clothing was paramount among the lady's maid's responsibilities, and demanded meticulous attention to detail.

During the winter and in wet weather, dresses needed to be carefully examined after they were taken off, and any mud removed. If a skirt or coat was made of tweed or some other woollen material it could be laid on a table and brushed all over, but for lighter fabrics it was better to beat them lightly with a handkerchief or thin cloth. Silk dresses were never brushed, but rubbed with a piece of merino, or other soft material of a similar colour, kept for the purpose.

A lady's maid was often required to be a dressmaker, at least for ordinary day wear. Evening gowns and the like were created for the mistress of the house and her daughters by professional dressmakers.

Summer dresses of barège (a sheer woven fabric of silk or cotton and wool), muslin, mohair, and other light materials, simply required shaking. If creased, muslin needed to be ironed.

Headgear needed equally careful attention. 'The bonnet,' says Mrs Beeton, 'should be dusted with a light feather plume, in order to remove every particle of dust ... Velvet bonnets, and other velvet articles of dress, should be cleaned with a soft brush. If the flowers with which the bonnet is decorated have been crushed or displaced, or the leaves tumbled, they should be raised and readjusted by means of flower-pliers. If feathers have suffered from damp, they should be laid near the fire for a few minutes, and restored to their natural state by the hand or a soft brush.'

∞ THE CLEANEST LINEN ∞

Removing unwanted stains and marks from delicate fabrics was a requisite skill for a good lady's maid. Some popular 19th-century treatments included:

Ink-spots – Pour on a few drops of hot water immediately after staining, then dip immediately in cold water.

Fruit stains on linen – Rub each side of the fabric with yellow soap, tie up a piece of pearlash (potassium carbonate) into the item and soak well or boil. Rinse and expose the stain to sun and air.

Grease stains – Pour some turpentine over the mark and rub until dry with a piece of clean flannel. Repeat as necessary, brush well and hang in the open air to remove the smell.

Wine stains on linen – Hold in boiling milk.

MAINTAINING LAMPS AND CANDLES IS A DAILY DUTY

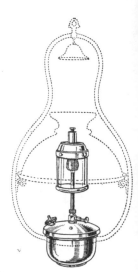

Dozens – if not hundreds – of lamps and candles were necessary to light a large country house, and their daily servicing was the specialist job of lamp- and candlemen.

Before the installation of electricity, a lamp room would be set aside for storing and maintaining candles, candlesticks and a plethora of oil lamps. Each morning, lamps would be collected from the rooms of the house and their wicks checked. If necessary these were trimmed, using scissors with blades fitted with circular lips to catch the burnt wick and prevent it from falling back into the lamp. From the 1850s, both colza oil (selected for its purity and pleasant smell) and paraffin were used for lamps and stored in large tanks in the lamp room.

Each lamp needed to be taken apart at least once a week to allow the sooty glass to be cleaned and the wicks to be trimmed. *Cassell's Household Guide* recommended the following routine: 'The works of oil lamps of every description should be soaked in hot water and soda, and rubbed perfectly dry whilst hot with a soft rag, and afterwards polished with a plate-leather. In trimming the cotton wicks,' it said, 'the greatest evenness is requisite.'

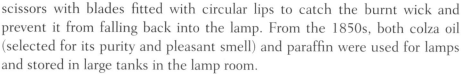

At dusk, lamps in the upstairs reception rooms were lit by footmen, while those of the cellars, basement and corridors were attended to by the hallboy. The candle boy was responsible for the rest of the house. For lighting, tapers were used.

Wicks needed to be treated with great care to make them last as long as possible. 'The wicks of paraffin lamps,' says the Guide, 'should only be dusted until the charred portions are removed. By this means a wick one-third of a yard in length lasts several months.' To prevent lamps from smoking, soaking the wicks in strong vinegar then allowing them to dry was an effective remedy. Metal globes and stands had to be rubbed over and polished if necessary. The wicks of candles also needed daily attention, while candlesticks needed to be carefully scraped in the lamp room to remove all remnants of dripped wax before being polished.

THE HOUSE MUST BE SET IN ORDER BEFORE STAFF RETIRE TO BED

A strict routine was essential every night to make sure the house was not only prepared for the following day but also secure.

Following dinner, tables were cleared by the footmen. Once the washing up was done, both cook and housekeeper needed to check that all was in order before going to bed, with all dishes and cutlery put away, although making sure that all preserves were properly stored was the housekeeper's duty. Meanwhile, the scullery and lower housemaids might still be at work blacking the kitchen range and laying fires while the lower footman or hallboy were doing jobs such as filling coal scuttles and attending to lamps and candles.

For the valet, the day could not end until after his master had gone to bed and any clothes were brushed by him personally or taken to the footman for attention. Similarly, the lady's maid needed to ensure that evening dresses were properly hung up and any laundry removed for washing.

Last to retire was the butler. It was his duty to see that all the glasses and silver (the 'plate') were washed and dried and put away in the pantry, the plate being kept under lock and key. His final task was to see that the doors were locked and shutters drawn, and that any fires that had been alight during the evening had died down to a safe level.

If they had days off, lower servants had to be back in the house by 10.00pm – or at least at an hour prescribed by the butler or housekeeper.

CHAPTER 3

THE COUNTRY HOUSE KITCHEN

*T*he kitchen of the country house, together with the many rooms attached to it with specific purposes from pastry making to food storage, was a hive of activity vital to the sustenance of both family and staff. In the earliest country houses the kitchen had a central position in the building but, as residences became larger and more complex in the late 17th century, it became separated from the main house and was provided with its own quarters. By the Victorian era, with the burgeoning of technology, it had become what has been described as 'a complicated laboratory' with a vast range of equipment, all of which needed to be employed, cleaned and cared for by the team of kitchen staff, from the cook or chef to the lowly kitchen and scullery maids. In every kitchen almost the entire day was taken up with preparing meals, either for immediate consumption or for grand events in the social calendar.

As far as possible, everything cooked in the country house kitchen had its origins on the estate, and menus were governed to a considerable extent by the seasons, although a great deal of food was preserved in some way for winter use. Where there were heated orangeries and greenhouses, exotic ingredients such as oranges and pineapples could be supplied, and were considered great luxuries. But whatever was cooked it was essential that the kitchen remained unobtrusive. It needed to run smoothly but, above all, be undetectable by sight, sound or smell by the family and their guests.

THE KITCHEN SHOULD NOT BE MADE OBVIOUS BY ITS ODOURS

A maxim dating from the early 20th century, when country house design brought kitchen and dining room into closer proximity, despite the kitchen having its own chimneys and vents.

In early country houses, keeping the kitchen separate from the rest of the house was as much a precaution against fire as anything else, but by the 1680s the trend was to remove the kitchen from the main building into a separate pavilion, which was sometimes joined to the house, as at Harleyford in Buckinghamshire, by an underground tunnel. This may have been inconvenient for serving meals but was most satisfactory in removing odours.

The use of dustbins was specifically discouraged for their tendency to make a kitchen smell. Where pigs were kept on the estate, food waste was placed in pig bins for feeding these animals.

By Victorian times the servants' wing of the house had become so extensive that the kitchen was often far removed from the living quarters. If placed on the same floor, it was usually connected to the dining room by a long passageway deliberately designed to be devoid of any staircases up which odours could travel. Even so, the cook needed to keep cooking smells to a minimum by ensuring that grease was not allowed to burn on the stove or in the oven. One trick was to put a crust of bread into the water in which green vegetables were boiled. After being drained, the cooking water was taken outside into the garden for disposal.

THE MAN COOK IS GENERALLY A FOREIGNER

While most 'ordinary' country houses employed women cooks, French chefs were kept by the wealthiest and most prestigious families.

If the man cook was an Englishman, it was said in the 19th century that he needed to 'possess a peculiar tact in manufacturing many fashionable foreign delicacies, or of introducing certain seasonings and flavours to his dishes which render them more inviting to the palate of his employer, than those produced by the simply healthful modes of modern English cooks.' When a man cook was employed he would often have his own room adjacent to the kitchen where he could compose menus and refer to his collection of recipes in quiet and comfort.

Such was the status of the man cook that women aspiring to be heads of kitchens in the largest establishments needed to have experience of working as a man cook's assistant. In 1777, for example, a female cook seeking employment was careful to stress that she had been 'brought up under a man cook'. But even when she was engaged to head the kitchen, the female cook always rated below the housekeeper in rank.

Whether male or female, the cook needed experience and knowhow. In *The Servants Book of Knowledge* of 1773, Anthony Heasel stressed the importance, particularly for women, of the spheres of 'provisions in general, and also of the most proper methods used in dressing them'. Among the latter would have been the many French ways of preparing, cooking and serving dishes of all kinds.

All head cooks held sway over the kitchen staff and many were as irascible as they were tyrannical. It was not unusual for them to quit their posts in high dudgeon without notice, so it was always in the mistress's interests to keep them as sweet as possible.

FOR THE COOK, AN HOUR LOST IN THE MORNING WILL KEEP HER TOILING ALL DAY

Wise words for country house cooks, who would need to be extremely well organized and timely in everything they did, especially when the house was full of visitors.

Although her assistants would rise before her, the cook needed to be up by 7.00 in the morning at the latest to ensure that everything was in order before staff breakfast. Early morning tea was made and brought to her by the kitchen maid. By 7.30 she needed to be ready to receive deliveries from the gardener and local tradesmen. Depending on the produce available, the cook would finalize the day's menus, which she would present to the mistress of the house following family breakfast, then adjust according to activities organized for the day and the needs of any guests. These meetings were also an opportunity to discuss large dinners, ball suppers and the like in advance.

Despite the appearance of cookery books such as those written by Eliza Acton and Mrs Beeton, most country house cooks rarely consulted recipes but carried the knowledge of everything they cooked – and the quantities required – in their heads as a result of years of experience.

Following the preparation and serving of lunch, cake baking, both for tea and for the following day (or further ahead for fruit cakes, which needed time to mature) was an afternoon activity, as were time-consuming activities such as making gelatine. However, serving dinner on time, and to the standard required in a great country

◅ An Autumn Dinner ▻

A November dinner suggested by Mrs Beeton, containing dishes that might well have been served to family and guests at the height of the country house season:

Ox-tail Soup   Soup à la Jardinière

Turbot and Lobster Sauce   Crimped Cod and Oyster Sauce

Stewed Eels   Soles à la Normandie

Pike and Cream Sauce   Fried Filleted Soles

Filets de Boeuf à la Jardinière

Croquettes of Game aux Champignons

Chicken Cutlets   Mutton Cutlets and Tomato Sauce

Lobster Rissoles   Oyster Patties

Partridge aux Fines Herbes   Larded Sweetbreads

Roast Beef   Poulets aux Cressons

Haunch of Mutton   Roast Turkey

Boiled Turkey and Celery Sauce   Ham

Grouse   Pheasants   Hare

Salad Artichokes   Stewed Celery

Italian Cream   Charlotte aux Pommes   Compôte of Pears

Croûtes Madrées aux Fruits   Pastry   Punch Jelly

Iced Pudding

Desserts and Ices

house, was the cook's major preoccupation. Not only were there four or more courses to be prepared, but there might be six or eight different choices for each course (see 'An Autumn Dinner' box). The height of activity was from about 6.00 in the evening to the serving of the last course of dinner at 9.00 or later. Only when everything had been cleared and the kitchen cleaned and made ready for the next day could the cook retire to bed.

THE STOCKPOT IS THE BASIS OF THE KITCHEN

Stockpots, kept constantly on the boil, were a starting point for almost everything savoury created in the country house kitchen, from soups to gravies and glazes.

The traditional stockpot was a large, lidded pot, sometimes with a tap at the base for releasing the liquid. In a large kitchen there would be two stockpots – one brown, one white – for brown and white soups and sauces. Bones always formed the basis of the stock: roasted beef bones for brown stock, veal bones and chicken carcasses for white. Other ingredients included meat of various kinds, onions (with the skins left on for brown stock), plus chopped carrots, celery, leeks or other vegetables to hand. Herbs such as bay, parsley and thyme would also be added, plus pepper and, if desired, cloves and mace.

Skimming the stock of fat while it was bubbling gently was a duty of the kitchen maids. At around 9.00 in the evening the stockpots would be taken off the range and the contents strained off through hair sieves to remove every possible particle of grease. Next morning they would be brought back to the boil and more meat and vegetables would be added.

For making clear soups, cloudy stock was clarified by adding eggshells, bringing the mixture to the boil and skimming or straining the liquid through a hair sieve or piece of muslin. Very strong clear stock was sometimes boiled right down to make a glaze, which was poured into skins (intestines, as for sausage making), then hung up to cool and harden. To use the glaze, a little was cut off and added to gravies for extra flavour.

∾ BEST STOCKS ∾

Tips for stock making from the Victorian kitchen, most of which are still totally valid today:

• *Allow a quart of water to each pound of meat.*

• *Don't add too many vegetables – they make the soup cloudy – and avoid turnips. (In the days before fridges these were also wise precautions against the growth of microorganisms.)*

• *Skim the pot as soon as it comes to the boil and before you add the vegetables.*

• *Boil constantly for five or six hours for a perfect stock.*

• *A partridge or pigeon makes a good alternative finish to beef or mutton.*

COPPER POTS ARE KITCHEN ESSENTIALS

The durability and reliability of copper pots, superb at conducting heat, made them vital equipment in the country house kitchen, where they could number in the hundreds.

All manner of kitchen utensils were made of copper. A typical selection, from an 1834 inventory at Erddig in North Wales included: 3 copper fish kettles, 3 copper preserving pans, 2 copper pots, 12 copper moulds, 3 copper frying pans, 24 copper cups and 25 copper stewpans. Then, as now, copper bowls were the very best for whisking egg whites.

Copper in the kitchen was only totally safe to cook in when coated inside with a thin layer of tin, which prevented acidic foods interacting with the copper itself. Without such a precaution there was a risk of verdigris poisoning, which resulted from the contamination of food with the copper oxides. For this reason the cook and her underlings all needed to keep a close eye on the scrupulous cleanliness of unlined copper utensils and the linings of coated ones. The latter could be sent for re-tinning if necessary, as the linings were soft and easily damaged by the careless use of metal utensils or harsh cleaning.

'Coppers' were large vessels or boilers in which water was heated in the scullery. These were regularly used for scalding cloths and, in the largest houses, for cooking vegetables.

To keep them gleaming, copper pans needed meticulous daily cleaning. For this the scullery maid used rotten-stone, a very fine abrasive powder, mixed with soft soap and oil of turpentine to a stiff putty.

VEGETABLES REQUIRE CAREFUL STORAGE AND TREATMENT

In country house larders vegetables – largely grown on the estate – were stored for use all year round. Or there might be a separate storeroom near the kitchen and scullery, convenient for preparation.

Individual types of vegetables were stored in stone boxes divided into compartments, either in the vegetable store or in the 'wet' larder (which was also the store for meat and fish if those were not assigned a separate larder). The boxes might be accommodated underneath sinks, or could be mounted on casters so that they could be easily moved around. Large quantities of root vegetables, such as carrots, were kept in the dark, covered in sand or matting. In smaller all-purpose larders also containing bread, cheese, meat and fish, vegetables would be laid on slate or marble slabs and covered with muslin or, in hot weather, wet rushes.

For winter use vegetables including onions, radishes, beetroot, cauliflowers and cabbages were pickled in vinegar. Others, notably tomatoes and marrows, were made into chutneys, while peas and beans were preserved by being layered with salt in stoneware jars.

The preparation of vegetables was the job of the kitchen maids or, if these were few in number, it might be delegated to a scullery maid. Instructions such as these were typical of the 19th century and would have been a daily requisite: 'Take care to wash and cleanse vegetables thoroughly from dust, dirt and insects, – this requires great attention. Pick off all the outside leaves, [and] trim them nicely…'

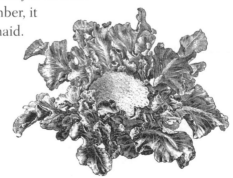

THE GREATEST POSSIBLE ATTENTION MUST BE PAID TO MAKING BUTTER

Because fine judgment is needed at every stage of the process. As it was most critical to ensure that the butter would set or 'come', the operation was often accompanied by the use of charms.

Milk for butter making was delivered from the home farm on the country house estate or provided by cows that the dairymaid milked herself (see Chapter 6). For a large establishment, butter would be made every day in summer when the cows were in full milk, and every two or three days at other times of year. The fresh milk was left in wide pans for 24 hours, then the cream, which had risen to the top, was separated off using a skimming dish.

After being left for two or three days to 'ripen', the cream was put into a churn then either rotated steadily or agitated with a plunger. Once the butter had 'come', the liquid buttermilk was poured off (to be fed to the pigs) and the butter transferred to shallow wooden tubs or bowls full of cold water – ideally pure spring water – in which it was squeezed and washed by hand. Several washings would be necessary before the water was clear, salt being added towards the end of this process. The final washing might be made in a butter worker, a shallow wooden tray fitted with rollers to mash the butter and to which salt could also be applied.

Once made, the butter was shaped with wooden pats known as butlets or 'Scotch hands', or pressed into a variety of fancy moulds.

Since Satan or witches were thought to prevent butter from 'coming', to ward off their ill effects dairymaids would throw silver sixpences into the churn, hang stones with holes in them in the dairy and strew the floor with protective herbs including betony, clover, dill and St John's wort.

⚬ BETTER BUTTER ⚬

Some hints and tips for the dairymaid:

Hands must be clean and all utensils sterilized by scalding them with boiling water.

Avoid rapid motion of the churn – it will force too much air into the butter and make it swell.

Any spilt milk should be immediately cleaned up with boiling water.

If butter refuses to come, wrap the cream in a calico cloth and leave it to drip for several days to make cream cheese.

To prevent summer-made butter from going rancid, put it in a stoneware jar on a layer of crushed salt and top with more salt.

A BREAD TROUGH SHOULD NEVER BE WASHED

Vital advice, needed to keep the trough impregnated with fermented yeast and to ensure that the dough would rise reliably time after time. When not in use, the kneading trough was used for storing baking tins, cloths and small quantities of flour.

The kneading trough in which bread dough was mixed was a wooden or slate container with thick, sloping sides and was, from medieval times, key equipment in the bakehouse – a separate room often adjoining the brewhouse (see Chapter 6). Instead of being washed after use the trough was rubbed clean with a ball of dough.

In a large house bread was baked every day, in smaller ones several times a week, depending on the number of occupants. At Ingatestone Hall in Essex in the mid-16th century, three kinds of bread were made – fine white bread or 'manchet', and two types of wholemeal bread, the one intended for the servants being the coarser.

Bread making began in the evening, when the flour was mixed with water and yeast, kneaded and left to rise overnight in the baking trough, covered with a wooden lid and warm sacks. The top of the lid, usually removable, was handy for shaping and moulding the dough before its final rising or 'proving'. A plain round loaf was finished by slashing a cross into the top, supposedly to keep out the Devil (but also to ensure perfect rising).

On a baking day, brick-built or brick-lined ovens would be lit in the morning, two or three hours before they were needed. The oldest country house oven was the beehive, shaped as its name suggests. Dry, fast-burning wood, such as furze and blackthorn faggots, which would heat the bricks thoroughly to a high temperature, were essential. The shaped dough was traditionally inserted into the oven on a peel – a pole with a wide blade at one end, but only after the ashes and embers had been removed from the oven base and the cavity cleaned with a damp cloth.

The bakehouse was dry and warm, and thus an ideal place to store flour, salt, buttermilk and yeast. The latter was kept in an earthenware jar with a cover over the top with a quill pushed through it to allow any gas from the fermenting yeast to escape. A dipper or ladle for carrying buttermilk or water was often fitted into the yeast jar. With the introduction of kitchen ranges, old-style bakehouses, such as can still be seen at Lanhydrock in Cornwall, became less common, and bread was made in the kitchen or an attached baking scullery.

MEAT MUST BE TURNED WHILE ROASTING OVER A BRISK FIRE

Useful advice from the country house kitchen before the advent of the kitchen range, when meat was roasted on an open fire.

Before the 18th century, meat was simply roasted in the medieval manner by putting it on a spit and turning it before a log fire, which in a large house might be raised up on a hearth within the fireplace. The spit was turned by hand, or sometimes even by a dog running on a kind of treadmill. For human servants it was hard, hot and thirsty work, and was superseded, from the 16th century, by a clockwork 'jack' driven by weights. Underneath the spit a lipped

∽ A PERFECT ROAST ∽

The Complete Servant of 1625 also included these tips for spit roasting:

Great care must be taken when putting meat onto the spit so that the prime part of the meat is not damaged.

For beef: a large joint should be kept a good distance from the fire at first, then brought nearer.

To prevent fat parts of the meat becoming scorched, tie kitchen paper over them. Do not skewer these on.

Allow a full quarter of an hour to a pound for roasting.

Pork must be basted with salt and water and, in particular, needs to be thoroughly done compared with beef and lamb. Its skin must be scored with a sharp knife after it has been by the fire for some time to 'make it eat better' – that is, produce crackling.

tray or dripping pan was placed to catch the fat and cooking juices, which could be ladled back over the meat to baste it.

A fire clear at the bottom, glowing and burning briskly, was essential for a good roast. The Regency guide, *The Complete Servant* also said that: 'The ashes should be taken up, and the hearth made quite clean, before you begin to roast. If the fire require to be stirred during the operation' it added 'the dripping pan must be drawn back, so that then, and at all times, it may be kept clean from cinders and dust. Hot cinders, or live coals dropping into the pan,' it warned, 'make the dripping rank and spoil it for basting.'

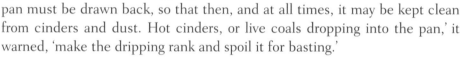

Beef, always a favourite meat in Britain, demanded special care. If it had been hung for some time it would need to be trimmed, soaked in salt water and dredged with flour before cooking, and basted continually with dripping while roasting. It was expected to lose about a third of its weight during cooking.

A WHOLE FISH IS BEST COOKED IN A KETTLE

Fish kettles were essentials of the country house cook's batterie de cuisine; *they were also used to cook shellfish ranging from lobsters to cockles.*

In a kettle, fish was usually cooked in a *court bouillon* – water with lemon or vinegar and herbs added, although the angler Izaak Walton recommended stale beer and a flavouring of horseradish in addition to the usual rosemary, thyme and winter savory. For trout, he said: 'Let your liquor boil up to the height before you put in the fish; and then if there be many, put them in one by one, that they may not cool the liquor as to make it fall.'

Before being simmered, fish was kept in the same larder as meat on a purpose-made slab where it could be immersed in a continuous stream of cold water. Alternatively it might be laid on sacks filled with ice and placed in zinc-lined cupboard-like containers. Near the coast, or at houses with extensive fishing, there would be a separate fish larder. Here, fish might even be kept alive, hung in a net filled with wet moss and fed with bread and milk in order, it was said, to maximize its flavour.

Away from coastal areas, eating the freshest fish from the estate was greatly preferred to sea-fish that might be over a week old by the time it reached the kitchen, as J.C. Loudon wrote in 1834: 'Instead ... we had a dish of the finest carp or tench I ever met with, or probably a jack, or eels, each taken from the stew ponds immediately before dinner, and thus eaten in the highest perfection.' Even the garnish was home grown – watercress for perch and wild thyme for trout. However freshwater fish often tasted distinctly muddy, even after the favoured treatment – which was to put a hunk of bread into the cavity left after the stomach had been removed.

Apart from the 'regular' fish kettle, well-equipped kitchens also had at least one turbot kettle. If served whole for a grand dinner the fish might be presented as Mrs Agnes Marshall recommended for *Turbot à la Chambord*, garnished with 'cooked crayfish, blanched and bearded oysters, cooked sliced lobster, button mushrooms, and whole truffles arranged around the edge of the turbot and some large slices of truffle in the centre'. A béchamel sauce made from the fish liquor made the finishing touch.

The original fish kettles were made of iron, which was succeeded by brass. This was followed, from the mid-18th century, by copper, which became the kitchen standard. For poaching a whole fish the kettle would be set on iron trivets placed on the stove.

HANG GAME IN A CURRENT OF DRY AIR

Houses on substantial estates were fitted with separate game larders away from the main house, where perfect ventilation could be ensured and unpleasant odours minimized.

The ideal game larder had large windows and plenty of air vents, with a cool slate and/or stone floor. For hanging whole carcasses, such as venison, it had rails or substantial wooden beams, while for hanging game birds, and smaller items such as rabbits and hares, a tiered iron frame was suspended from the ceiling. Chopping slabs were provided for butchering. By the turn of the 20th century, small cooling plants – pipes filled with brine cooled with compressed carbon dioxide – were being installed for long-term storage.

Before being cooked and eaten, venison was hung for at least two weeks, after which it was washed in milk and water before being dried with clean cloths. Any parts of the meat likely to be attacked by flies were rubbed with ground ginger or pepper. Good venison would have fat that was thick, clear and bright, not green or black. Hares were hung for a similar period before being 'jugged' with onions, herbs and possibly port added, as well as some of the animal's own blood.

Good game birds needed to be fat. If stale, the skin would peel off

Elegant but efficient, the game larder designed by Humphrey Repton for Uppark in West Sussex in the early 19th century is a perfect example. Octagonal in shape, it is divided into two 'rooms': one for smaller game, the other for carcasses. Additional cool storage space is supplied by marble shelves placed in alcoves.

when rubbed with the finger. Tough old geese were betrayed by their red feet and yellow, skinny fat. The preparation of game birds, which included quails, snipe, woodcock, plovers and even larks, was a time-consuming business. As well as having to be plucked and gutted they needed to be stuffed (if large), trussed and, if likely to be dry, larded by having strips of fat literally sewn into the breast with a larding needle.

A COOK'S SAUCES MARK THE CREDIT OF HER KITCHEN

Not only that, but each sauce needed to have its own character, fitted to the dish it accompanied. By the late 18th century sauce was, in grand houses, served separately in a sauce boat.

Making sauces was a preoccupation of the country house kitchen, particularly when catering for grand occasions, and was usually the duty of the kitchen maid. For grand cooking between the 15th and 17th centuries, the basis of a good sauce was the 'cullis', used for both thickening and flavouring. It was made with butter and breadcrumbs, and could be flavoured with truffles, crayfish or champagne. For simpler everyday dishes thinner sauces made with butter and a little flour plus cream or stock were used, flavoured with parsley, anchovies, shrimps and the like. These eventually became the norm but evoked the scorn of 'foodies' of the times for their blandness and tendency to lumpiness.

By 1800, thanks to the influence of French chefs such as Antonin Carême, three essential

Until the 1600s most sauces were strongly spiced with nutmeg and cloves. Often they would also contain ingredients such as raisins as well as lemon juice and vinegar or verjuice, made from unripe grapes.

sauces had become established in the cook's repertoire: the béchamel, the velouté, made with a light stock, and the éspagnole, based on a brown stock. As Mrs Beeton prescribed several decades later, each needed to 'possess a decided character; and whether sharp or sweet, savoury or plain, they should carry their names in a distinct manner ...' This precept also applied to sauces such as mint, apple, gooseberry and onion, as well as mayonnaise, which by this time was popular for cold dishes.

GOOD GRAVY

To accompany roast meat good gravy was essential. It could be made in two ways:

Using breadcrumb thickening – *A recipe from* Enquire Within *of 1894 typifies the cullis method: 'Three onions sliced, and fried in butter to a nice brown; toast a large thin slice of bread until quite hard and of a deep brown. Take these, with any piece of meat, bone &c., and some herbs and set them on the fire, with water according to judgement, and stew down until a rich gravy is produced. Season, strain and keep cool.'*

Using a roux – *From Cassell's Household Guide, using gravy 'produced by the cooking of meat'. This is 'thickened, without being boiled down, by the addition of various materials, such a roux of flour and butter...'*
Alternative thickenings included the yolks of hard-boiled eggs and chestnuts roasted and crushed into a powder. Good stock or white or red wine were all recommended additions.

THE TASK OF KEEPING FOOD HOT DESERVES CONSIDERABLE ATTENTION

The greater the distance between kitchen and dining room, the more difficult it was to keep food hot for serving. The secret of success lay in both equipment and timing.

Advising cooks on the importance of hot food, *The Complete Servant* stressed the need for planning and timing: 'The clock must be consulted, and the different articles prepared and laid to the fire, in succession, according to the times they will take, that all may be ready in due time. A scene of activity now commences,' it says, 'in which you must necessarily be cool, collected and attentive. Have an eye to the roast meat, and an ear to the boils, and let your thoughts continually recur to the rudiments of your art, which at this moment must be called into practical requisition. You will endeavour that every kind of vegetable, and of sauce, be made to keep pace with the dishes to which they respectively belong – so that all may go upstairs *smoking hot* together, and in due order.'

For breakfasts and informal suppers, food could be kept hot in a chafing dish. Traditionally made of silver, it consisted of an outer container filled with water into which an inner pan containing the food was placed. The whole was heated from below by an oil or spirit lamp.

For food to have a chance of being a good temperature when it reached the table it needed to be piping hot when it left the kitchen. All hot dishes were kept simmering on the stove or left in the oven before being lifted onto thoroughly heated serving dishes, which were then quickly covered with large metal domes, also heated. Plates, too, needed to be heated before being sent to the dining room. Soup was sent up in a hot tureen to be ladled into hot plates in the dining room.

To ensure perfect timing, cook and staff had to work in harmony. The butler was the essential go-between in this and ideally advised, to the minute, when food should leave the kitchen.

Mixing the salad is the duty of the butler

Although the salad was usually made by the cook or one of her assistants, the final touches, added just before serving to prevent limpness, were the butler's responsibility.

Salads were supper dishes that, at their simplest, consisted of mixed green leaves such as lettuce, watercress, mustard and cress, but possibly also corn salad and sorrel. The country house salad might also contain young nettle and dandelion leaves, and flowers such as nasturtiums. In winter, cabbage (commonly pickled) and shredded Brussels sprouts were used, as well as celery and chicory. Other common ingredients included beetroot, cucumber, radishes, spring onions and potatoes. Cooked carrots, cauliflower and French beans might be used too.

For a more substantial dish, meat was added to a salad, as in the popular

Tomatoes were not a standard salad ingredient until the 19th century. Originally regarded with some suspicion, they were more likely to be served baked or as a flavouring for soups and gravies. Salad flavourings included chervil, mint, parsley, tarragon, sage and garlic.

Victorian 'Salmagundi' – a 'composed' salad that might consist of a ring of blanched lettuce hearts surrounding decoratively arranged circles of beetroot, egg yolks, watercress, egg whites, chopped cold meat and pickled cabbage. Seafood might be used instead of meat, and nuts could also be added. Among the many 'fancy salads' devised by Mrs Agnes B. Marshall were an Italian salad of cooked potatoes, cucumber, boiled cauliflower, artichoke hearts and Brussels sprouts in a dressing flavoured with tarragon, chervil and chopped shallots, garnished with anchovy-stuffed olives.

⋘ THE PERFECT DRESSING ⋙

Enquire Within amused and informed its readers with Sydney Smith's 'Salad Mixture in Verse'. The 'oil of Lucca' is Italian olive oil.

Two large potatoes, passed through kitchen sieve,
Unwonted softness to the salad give;
Of mordant mustard add a single spoon –
Distrust the condiment which bites so soon;
But deem it not, thou man of herbs, a fault
To add a double quantity of salt;
Three time the spoon with oil of Lucca crown,
And once with vinegar procured from town.
True flavour needs it, and your poet begs
The pounded yellow of two well-boiled eggs;
Let onion atoms lurk within the bowl,
And, scarce suspected, animate the whole;
And lastly, on the favoured compound toss
A magic teaspoon of anchovy sauce;
Then, though green turtle fail, though venison's tough,
And ham and turkey be not boiled enough,
Serenely full, the epicure may say, –
'Fate cannot harm me – I have dined today.'

THE BEST JELLY HAS A PERFECT CLARITY

Jellies were great favourites in the country house but demanded a great deal of time and effort, as gelatine had to be made from calves' feet. From the 18th century, isinglass became a reasonable, time-saving substitute.

A jelly had to be planned well in advance. First a stock had to be made by boiling a couple of calves' feet in water for six or seven hours. The reduced liquor was then strained and cooled, and any fat removed from the top. Mrs Beeton then recommended pouring a little warm water over the top to remove any last traces of fat. The clear jelly was carefully separated from the sediment in the bottom of the container.

Isinglass, a form of the protein collagen, was a quick substitute for calves' foot gelatine. Until 1795, when a cheaper and more readily available version was made from cod bladders by the Scots inventor William Murdoch, it was made from Beluga sturgeon. Prepared gelatine became commercially available during the 19th century.

Next, a quart of the stock was put into a saucepan with 6oz (170g) loaf sugar, plus the shells and well-whisked whites of five eggs. It was gently brought to the boil – but never stirred – and allowed to bubble for 10 minutes before a teacupful of cold water was thrown in. After another five minutes' simmering the pan was removed from the stove and left in a warm place for half an hour. Finally the mixture was passed through a jelly bag made, so Mrs

Beeton recommended, of 'the very stout flannel called double-mill, used for ironing blankets'. If it did not come through clear then it needed to be put through the bag a second time. The resulting stock, she said, was 'the foundation of all *really good* jellies, which may be made in varied in innumerable ways, by colouring and flavouring with liqueurs, and by moulding it with fresh and preserved fruits'.

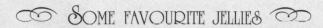

SOME FAVOURITE JELLIES

Popular set desserts from the Victorian country house. These were made in metal moulds of varying sizes and designs, many of which are now prized by collectors of kitchen antiques.

Lemon jelly – Lemon syrup made with lemon juice and rind and set. Orange jelly was made in a similar way.

Bohemian cream – Puréed fruit with cream set with isinglass.

Coffee jelly – Strained coffee mixed with dissolved gelatine and a custard made from milk, sugar, eggs and vanilla.

Claret jelly – Claret fortified with brandy and set with gelatine. The dish was garnished with a compote of plums and whipped cream.

Sultana sponge jelly – Sultanas boiled with cinnamon, lemon peel and bayleaf, then strained and pulped. The strained liquor was warmed with gelatine and rum, brandy and Maraschino liqueur with vanilla added. When cool it was whipped to a sponge-like consistency before being put into a mould to set.

EVERY DISH SHOULD BE SENT TO THE TABLE PROPERLY GARNISHED

By the mid-19th century, garnishing dishes was not only expected but had become a culinary art that took up considerable time.

Such were the demands of good dining that the simple sprig of parsley or sprinkling of chopped herb was an inadequate garnish. For cold dishes, elaborately shaped decorations were made by carving root vegetables such as carrots and turnips into floral shapes and leaving them in cold water to 'blossom', scooping cucumber flesh into 'peas' or dicing artichoke hearts. Lemon and mushroom slices were popular, as were cooked prawns and shrimps, which might also decorate soup.

For hot dishes, garnishes might well be cooked. Mrs Agnes B. Marshall recommended garnishes such as cucumber flesh cut into olive shapes, boiled, then tossed in butter, lemon juice and parsley, or button onions blanched and braised. One of her more complex suggestions was a 'Reform Garnish' for cutlets or a braised fowl, which sounds almost like a dish in itself: 'Take Julienne shreds of French gherkin, truffle, button mushroom, cooked ox tongue or ham, and hard boiled white of egg, making in all half a pint; put between two plates with a little white stock or water and boil over boiling water.' To create small vegetable garnishes, cooks used special implements such as pea-cutters and melon ballers.

The most elaborate garnish of all was spun sugar, which involved boiling sugar to a perfect 'crack stage' caramel (extremely difficult without the aid of a sugar thermometer) then using implements such as forks and spoons to 'throw' it over a rolling pin and draw it into fine threads.

THE WET LARDER MUST CONTAIN EQUIPMENT FOR SALTING

Before refrigeration, preserving meat was of primary concern in the country house kitchen, and was carried out in the wet larder in which it was traditionally stored.

As well as meat intended for long keeping, even fresh meat was rubbed with salt to help prevent it from becoming tainted, but before this it would always need scrutinizing to ensure that it was not afflicted with 'fly strike': that is, with eggs laid by houseflies. Any suspicious lumps of fat – especially gelatinous 'kernels' – were cut off.

For longer preservation, a mixture of salt and saltpetre (potassium nitrate) was used, kept in a powdering tub, and meat was placed on wooden trestles or on a salting stone or shallow sink in the larder. In addition, troughs for salting might be placed around the larder, supported on bricks. Deep troughs were for long-term storage, shallow ones for meat intended for more immediate

---

MAKING HAM AND BACON

As well as being salted, ham and bacon needed to be smoked, usually in a separate smoking room.

For ham – Rub the pork with a little salt and leave it overnight. Next day, boil together half a pound of bay (sea) salt, half a pound of common salt and half a pound of moist sugar with a quart of stale beer. Pour it over the meat and baste and turn it for three weeks. Dry the ham over a fire of wood, sawdust and peat.

For bacon – Treat the meat as for ham, then roll it in bran before hanging it up to smoke. Or enclose it in a coarse cloth, sewing it in securely, then hang it for a week in a baker's chimney.

use. Both were filled with a pickling solution, which typically contained salt and saltpetre plus sugar and flavourings such as bay leaves. For economy, pipes from the bases of the troughs to containers beneath allowed the brine to be drained off, re-boiled and re-used.

An alternative to the troughs were lead-lined wooden tubs or large earthenware crocks. Meat could be left in them for up to 12 months, being periodically topped up with salt, but two to three weeks' steeping was more usual.

PRESERVES ARE THE HOUSEKEEPER'S PROVINCE

Making pickles, ketchups, chutneys, jams, jellies and marmalades, plus syrups, distilled waters, and even essential oils and cosmetics, were important duties for the housekeeper, carried out in the stillroom with the assistance of one or more kitchen maids.

All kinds of vegetables grown in country house gardens were pickled. Onions, beetroot and cabbage – both green and white – were favourites, as were green and red tomatoes. Unripe walnuts were also pickled in vinegar, as were fruits of all kinds, including plums, peaches and, if the estate boasted an orangery, lemons and oranges. Ketchups were made from everything from tomatoes (see box) to cucumbers, mushrooms and even oysters.

'Beverages' for medicinal use were made in the stillroom. For seltzer and soda waters, a forcing pump was used to add carbonic acid gas to the mixtures to make them fizz.

Vinegars were flavoured with anything from gooseberries, currants, horseradish, mint and chilli to elderflowers, roses and dried primrose petals. Curry powders were also mixed, as were other flavourings such a 'peas powder' made from dried mint and sage, celery seed and cayenne pepper or allspice. Sugar for jams and jellies needed to be broken up or 'nipped' ahead of use, but they were otherwise made in exactly the same way as today. Syrups were flavoured with orange or lemon peel, or used to preserve apples and other fruit, often with large amounts of brandy or other liqueur added.

∽ COUNTRY HOUSE KETCHUP ∾

A recipe from Tendring Hall in Suffolk from 1857, quoted by Florence White in her 1932 book Good Things in England. *The chilli vinegar would have been made in the stillroom and the sieving done by the stillroom maid. The final note emphasizes the fact that waste was frowned upon, although the safety of the suggested method is decidedly questionable.*

Ingredients: tomatoes, quite ripe; chilli vinegar; salt, garlic ½ oz; or shallots 1 oz to each quart.
Time: bake for about 15 minutes; boil for 15.

Method
1. Bake quite ripe tomatoes till they are perfectly soft.
2. Rub the pulp through a sieve.
3. Add as much chilli vinegar as will make a fairly thick cream.
4. Slice the garlic or shallot and boil all together for 15 minutes.
5. Take the scum off.
6. Strain it through a sieve to remove the garlic or shallot.
7. When cold, bottle and cork it well.

N.B. – If when the bottles are opened it is found to have fermented, put more salt to it and boil it up again. The thickness when finished should be that of very thick cream.

BREAKFAST TOAST MUST BE PREPARED IMMEDIATELY BEFORE SERVING

Before the days of electric toasters, preparing toast each morning was a tricky task that demanded perfect timing. Once done, it had to be handed immediately to a footman to be carried to the dining or breakfast room.

Toast was an essential addition to the country house breakfast buffet laid out on a sideboard, which also contained cold joints and cuts of meat and brawn, plus hot dishes such as kidneys, hashed mutton and kedgeree. Thin slices of toast were also served topped with strained, boiled bone marrow, sprinkled with pepper and parsley.

Bread a day or two old was ideal for toast. After being cut into even slices it was impaled on a toasting fork and browned in front of the fire – ideally glowing but not smoking embers. The difficult part was preventing it from burning. For as 'Wyvern' (Col. A.R. Kenny-Herbert) said in his 1894 book *Fifty Breakfasts*, toast is a 'very simple thing to be sure, yet how often is it maltreated, scorched outside, spongy within and flabby?' Diligence and concentration on the part of the cook or one of her assistants was vital, and slices might have to be taken on and off the toasting fork several times and turned to prevent them from curling.

Once done, Mrs Beeton recommended putting toast on a hot plate, then cutting good butter (quality was essential) into small pieces and placing them on the toast. The plate should then, she said, be put beside the fire until the butter was just beginning to melt, when it could be spread lightly over the toast. Finally, she says: 'Trim off all the crust and ragged edges, divide each round into 4 pieces, and send the toast quickly to table.'

For ease of toasting Mrs A.B. Marshall sold, by mail order, folding racks to hold toast, imported from America, which sold for 3 shillings and could be used on hotplates.

SAVOURIES ARE SERVED AT BOTH LARGE AND SMALL DINNERS

Especially popular in Victorian times – and with male diners – savouries were served after sweet dishes and before ices. At smaller dinners they might be prepared as an alternative to sweet dishes.

True to their name, savouries were well flavoured, and often served on toast, although they could contain sweet elements such as prunes and raisins. Marinated herring, anchovies and devilled shrimps were regularly employed. Popular recipes included baked herring roes, shrimp toast and scalloped lobsters. A typical topping, in an era when oysters were cheap and plentiful, was an oyster sauce made with fresh oysters flavoured with lemon juice, cayenne and salt. Savouries that involved setting ingredients such as caviar and lobster in a gelatine-base aspic, were often served in paper cases. Or small pastry cases or profiteroles might be filled with a cheese mixture.

Devilled muscatel raisins, deep fried then seasoned with salt, paprika (known as coralline pepper from its colour) and ground ginger were a variety of fruit-based savoury.

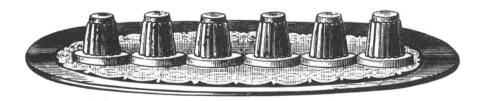

CHAPTER 4
A MATTER OF MANNERS

*O*beying the rules of etiquette was absolutely vital to the smooth running of the country house, not least to make sure that everyone was comfortable in their position and that the barriers between 'upstairs' and 'downstairs' remained intact. For guests, knowing the rules was even more vital, for it was incredibly easy to offend with the wrong dress, speech or action. Equally, servants needed to know how to address and care for everyone in the household, whether it was meeting guests at the station, serving food and wine at meals or running a bath. The good hostess would, of course, inform her guests ahead of a visit about any activities that would demand particular modes of dress, but it was incumbent on guests to know the form – and stick to it come what may. The same applied to servants such as ladies' maids and valets accompanying their masters and mistresses on country house visits. Discretion was also crucial. To discuss the business of the house with strangers was always unacceptable.

Throughout the house, orders of precedence prevailed and needed to be adhered to strictly. This not only involved knowing whose orders needed to be obeyed but also which parts of the house were out of bounds, even within the confines of the servants' quarters. For family and guests, rank and age needed to be taken into consideration, and for unmarried girls chaperones were always on hand, although for the determined there were means of escaping close supervision.

EVERYONE IN THE HOUSEHOLD MUST BE CORRECTLY ADDRESSED

It was as important for staff in the country house to be properly addressed by their employers as it was for staff to converse correctly with members of the household.

Forms of address depended largely on staff seniority. House steward, butler and valet were addressed by their surnames, while the housekeeper was always given the title of 'Mrs', even if she was unmarried. In similar vein, a cook was always 'Mrs' but if a 'man cook' was kept he was known as 'Monsieur' (since he was usually French). A governess was always 'Miss', as was a lady's maid, but the latter could also be addressed simply by her surname or first name – either was acceptable.

'Lower order' servants would address their superiors such as the butler, housekeeper and cook in the same way. For them, however, even their own names might not be acceptable for general use. They would either be called by their surnames or, if first names were used, often had to answer to names given to them by their employers. James and John were popular for footmen and Emma for housemaids. Emily, Jane and Mary were also

Servants of long standing were held in such regard that masters of the house might consider menservants their friends and kindly mistresses put themselves out to care for staff if they fell ill. Sir John Boileau, master of Ketteringham in Norfolk from 1836, declared that his servants 'were as much part of his family as his children'.

regularly used for female servants. Any unusual given name – even Ada or Marion – would commonly be banned as pretentious.

The conventions for using names epitomized the relationships between servants and their masters and mistresses. While they helped guard against over-confidentiality, employers were warned against adopting 'an arbitrary or haughty demeanour' and were advised instead to inspire confidence in their staff. Yet even a caring attitude needed to be distinguished from over-familiarity, which might breed idle gossip.

A LADY'S DRESS MUST BE ADAPTED TO CIRCUMSTANCES AND VARIED WITH DIFFERENT OCCASIONS

Being correctly dressed at all times was a priority for the mistress of the country house and her daughters, assisted by their ladies' maids. Several changes were required each day – and more if sporting activities such as hunting were enjoyed.

To start the day, the ladies of the house would wear simple morning dresses. If going out walking later in the morning, they might, in the Edwardian era, change into a walking skirt and blouse, possibly a *trotteuse* skirt that cleared the ground and was shorter at the back than the front. A matching coat would be worn, and possibly a waistcoat. Neat, smart footwear was a must.

For morning calls yet another change was required, but 'anything approaching evening dress,' warned Mrs Beeton, was 'very much out of place'. For a visit of condolence it was appropriate to dress in black. Any outing away

> *Changes of dress were essential for some female servants, too. Housemaids, for instance, wore print dresses for their morning work then, following luncheon, changed into dark ones, topped with frilly cap and white apron.*

from the confines of the house demanded a hat as well as outerwear of some kind such as a cape or coat and, in winter, a fur wrap. For afternoon wear, a tea gown was also appropriate but, said the magazine *The Lady's Realm*, it 'must be of good fabric' if it was to 'fall well'. It might be made of 'satin or silk, veiled with silk or lace'.

Dinner, or an occasion such as a ball (see also Chapter 5), required something altogether grander. Advising its readers on dress for autumn country house parties of the early 1900s, *The Lady's Realm* suggested 'the stiffest velvets and brocades of the Louis XVI period' with decoration such as family lace falling over the *décolletage*. For young women it recommended a satin gown lined with soft muslin and with a bodice 'softened by a chemisette of thin frills of Indian muslin or chiffon'.

WHEN RECEIVING CALLS, ETIQUETTE MUST BE GIVEN PROPER ATTENTION

As with paying so-called 'morning calls' – that is, calls made between luncheon and dinner – receiving visitors also demanded correct behaviour from both staff and the mistress of the house.

'There is no surer indication of the manner in which a household is conducted,' says *The Servants' Practical Guide* of 1880, 'than is conveyed in "answering the door".' The footman was instructed to open the hall door wide and to stand in the centre of the doorway before receiving cards from the visitors and, if the lady of the house is 'at home', (that is, not only in residence but prepared to receive visitors) usher the visitors towards the drawing room. At the drawing room door the footman opened the door wide and announced 'Mrs A.', 'Lady Emma F' or 'The Duke and Duchess of M', according to their position in society, correct wording being obligatory.

When visitors entered the room, counselled Mrs Beeton, occupations such as drawing, music or reading should be suspended, but if, she says, a lady is 'engaged with light needlework, and none other is appropriate in the drawing-room, it may not be, under some circumstances, inconsistent with good breeding to quietly continue it during conversation, particularly if the visit be protracted, or the visitors be gentlemen.'

For business calls, the etiquette was slightly different, with visitors being asked to take a seat in the hall. The footman, having ascertained the nature of the call, would take the visitor's card and either present it to his mistress on a salver or give it to her lady's maid for presentation. If he was taking the card to her himself he would correctly say: 'A person has called to see you, please ma'am, and is waiting in the hall.' Or, 'A lady wishes to know if you will see her for five minutes, and has sent up her card, if you please ma'am.'

When she judged that a visit had come to an end, the mistress of the house was advised to ring a bell for the footman then accompany her guests 'as far towards the door as the circumstances of your friendship seem to demand'.

FOR A COUNTRY HOUSE VISIT IT IS ALWAYS NECESSARY TO SELECT CORRECT AND SUITABLE CLOTHING

Being properly dressed for every occasion was essential country house etiquette for guests, and needed to take account of every activity from riding and walking to formal dining.

The fact that the country house 'season' took place in cooler months was reflected in the advice available for women. 'If she rides or hunts,' said one Edwardian manual, 'the girl must, of course, have the requisite kit. Otherwise,' it continued, 'she must have a tweed suit with a couple of woollen jumpers, a fur or other overcoat to wear over it in motoring, possibly a second tailor-made, and a simple day or a tea frock into which to slip when coming in from the hunt, or tramp, or motor run, to tea.' In addition low brogue shoes and woollen stockings were needed for traversing the moors,

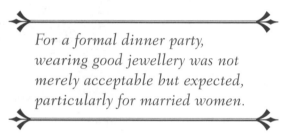

For a formal dinner party, wearing good jewellery was not merely acceptable but expected, particularly for married women.

plus scarves and 'close fitting little hats' for warmth. For later in the day two or three simple evening gowns were recommended, depending on the length of the visit.

For men, tweeds were essential for outdoor pursuits, in a weight fitted to the activity and the weather, and specifically tailored for activities such as shooting (see Chapter 5). Lighter wear was needed for tennis or croquet, and from the 1880s a blazer was acceptable for summer occasions out of doors. For more formal occasions a morning suit and dark waistcoat were needed for daytime wear and a tailcoat and white tie for dining. Trousers were loose and tubular cut and, from the mid-19th century, shoe length. Cane, gloves and hat were obligatory accessories.

UNTIL LUNCHEON, VISITORS AT A COUNTRY HOUSE SHOULD LOOK ABOUT AND AMUSE THEMSELVES

Polite country house visitors were usually expected to arrange their own morning entertainment unless their host and hostess had some pre-planned outing or activity arranged.

Walking in the gardens or countryside would have been considered totally acceptable pre-luncheon activities, as would a drive out in a carriage or, later, a motor car. Once cycling became both fashionable and acceptable for both men and women in the 1880s, cycle rides became another possible morning activity, and a good host and hostess might provide cycles for guests to use.

For motoring, which was the exclusive province of the wealthy until after World War II, the tailored suit was an ideal garment. Hats were a must and, for women, were tied with a fetching chiffon scarf over the top to keep them in place in open-top vehicles.

Cycling costumes were specifically designed for the activity; the women's magazine *Home Chat* of 1896 recommended those from Nicoll in London's Regent Street, a store patronized, it assured its readers, by the Royal Family. 'A soft, light cloth should be chosen,' it said, while 'a well-cut Norfolk jacket, a bright tie and a wide sailor hat with riband to match combines comfort and shade.' By 1904 *The Lady's Realm* said that 'the skirt for cycling should have an inverted pleat at the back, so that it falls on either side of the saddle ... Tweed or a hard wearing blue serge' were recommended as being 'always the most serviceable fabrics'.

THE ETIQUETTE OF SHOOTING IS MORE EASILY ACQUIRED THAN THE ART

Knowing how to behave – in every respect – was crucial to the success of a shooting party, particularly for country house guests, both male and female.

On arrival for a shooting party, good guests would take their guns and cartridge bag to the gun room for safe keeping or, for a shoot on a grouse moor, leave them in the care of the keeper. To prepare for a day's shooting of birds, including grouse, pheasant, ducks and partridges, both landowner and male guests or 'guns' would dress in 'uniform' tweeds, complete with Norfolk jackets whose full sleeves did not restrict the arms as the gun was raised, and a hat of some kind such as a deerstalker or tweed cap. Such clothes were essential for both warmth and camouflage. Before the shoot began, the correct host would ensure that everyone in the party knew the rules for the day, such as 'no ground game' or 'no woodcock'. Until it began, guns would be carried uncocked and cartridges secured in a bag carried over the shoulder.

Out in the field, the guns were placed at numbered stakes set in the ground at each site or 'drive' and took aim as the birds were driven overhead. The good host ensured that positions were rotated to give everyone a fair

⌒ ADVICE IN VERSE ⌒

In 1902 the sportsman and army officer Mark Beaufoy wrote a poem entitled 'A Father's Advice', which sums up the manners of the shoot:

If a sportsman true you'd be
Listen carefully to me …

Never, never let your gun
Pointed be at anyone.
That it may unloaded be
Matters not the least to me.

When a hedge or fence you cross
Though of time it cause a loss
From your gun the cartridge take
For the greater safety's sake.

If twixt you and neighbouring gun
Bird shall fly or beast may run
Let this maxim ere be thine
'Follow not across the line.'

Stops and beaters oft unseen
Lurk behind some leafy screen.
Calm and steady always be
'Never shoot where you can't see.'

You may kill or you may miss
But at all times think this:
'All the pheasants ever bred
Won't repay for one man dead.'

chance. However, it was considered poor form to pass across another man's sight, or to cross a neighbour's field to get to a wounded bird. Onlookers attending the shoot needed to keep quiet except while walking between drives.

At the shoot's end, it was considered vulgar to announce how many birds you had bagged unless specifically asked to do so by the host. House guests then paid the required fee to the gamekeeper (as agreed with their host) and took their guns to the gun room to be cleaned. If they wished to clean their own gun they would have needed to ask permission ahead of their visit.

IF COMMISSIONED TO TAKE CHARGE OF A LADY IN THE HUNTING FIELD A MAN MUST SACRIFICE HIS SPORTING INSTINCTS TO A CERTAIN EXTENT

Equally, so 'Madge of Truth' advised in her 1898 book, he must 'see her safe over the fences and give her a lead as circumstances may dictate'.

As with shooting, hunting demanded meticulous attention to manners by everyone involved. Anthony Trollope, describing the hunting exploits of Lizzie Greystock in *The Eustace Diamonds* in 1872, makes the same point, Lizzie being advised by Lord George to 'Follow me close, but not too close. When the men see that I am giving you a lead, they won't come between. If you hang back, I'll not go ahead. Just check your horse as he comes to his fences, and, if you can, see me over before you go at them.' And later in the piece Trollope tells us that he let her '… take the leap before he took it, knowing that, if there were misfortune, he might so best render help.'

Dressing correctly was vital. Even if entitled to wear 'the pink' (a red hunting jacket) on his own territory, no good guest would don anything but a black jacket, white breeches and a velvet or silk hat. For women the 'uniform' was a long black skirt, tight-fitting black

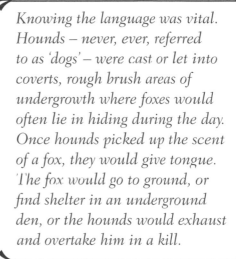

Knowing the language was vital. Hounds – never, ever, referred to as 'dogs' – were cast or let into coverts, rough brush areas of undergrowth where foxes would often lie in hiding during the day. Once hounds picked up the scent of a fox, they would give tongue. The fox would go to ground, or find shelter in an underground den, or the hounds would exhaust and overtake him in a kill.

jacket, stock and a hat complete with a veil or chiffon streamers.

In the field any man would be careful not to outrun or upstage the Master of Foxhounds and would be sure to give the groom who had charge of his hunter a generous tip.

ONCE MADE, AN ENGAGEMENT AT A BALL SHOULD NEVER BE BROKEN

One of the many rules of etiquette at a ball, where a lady would have her dance card filled by male partners. If breaking the 'engagement' was unavoidable, then it was polite for neither party to take part in that particular number.

'If, for instance,' one Victorian guide stated, 'two partners should claim one lady for the same quadrille or valse, the lady, having inadvertently engaged herself to both, should decline dancing with either, but should set the gentleman free to choose other partners.' Before requesting a dance with a lady to whom he had not been introduced, a polite gentleman would need to ask his host or hostess, or a member of the family, for a formal introduction. At the end of each dance a gentleman would be expected to offer his left arm to his partner and lead her to a seat near her chaperone. He would then withdraw as soon as her next partner arrived.

From start to finish, good manners at a ball were paramount, from the sending of timely invitations to calling on the host and hostess – or leaving a card at the very least – within two or three days afterwards.

Dancing at a country house ball was opened by the mistress of the house or by the master and the lady of highest rank. By the mid-19th century the quadrille was the dance of choice to begin a ball but, said Mrs Beeton, 'It will be well for the hostess, even if she be very partial to the amusement, and a graceful dancer, not to participate in it to any great extent, lest her lady guests should have occasion to complain of her monopoly of the gentlemen, and other causes of neglect.'

For dress, guidelines for late 19th-century young ladies recommended gowns 'of a light and gauzy kind, and of a length of skirt that enables the wearer to thread her way without impediment to herself and other dancers. Trains,' it said, 'are quite out of place in a ball-room, and even if carried over the arm, are simply an encumbrance.' For gentlemen the required garb was 'the ordinary black suit that constitutes full evening dress, with very open waistcoat, white necktie, and light lavender or white kid gloves.' A buttonhole bouquet of 'choice flowers' was considered acceptable.

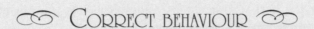

CORRECT BEHAVIOUR

More rules for balls, as expounded by Enquire Within:

- *Upon entering, first address the lady of the house, and after her, the nearest acquaintances you may recognize in the room.*

- *Avoid excess of jewellery.*

- *Do not select the same partner frequently.*

- *Never stare about you as if you were taking stock of those present.*

- *The host and hostess should look after their guests, and not confine their attentions. They should, in fact, attend chiefly to those who are least known in the room.*

COUNTRY HOUSE GUESTS WHO ARE ARRIVING BY TRAIN SHOULD BE MET AT THE STATION

And, furthermore, they should be supplied before their arrival with details of the train timetable. Meeting guests would be the duty of the second coachman or, in later eras, the chauffeur.

Once guests' arrival had been announced, the hostess would meet and greet them personally. She would, ahead of time, have stipulated whether women guests might be accompanied by their personal maids and men by their valet, chauffeur and, if shooting was on the agenda, their personal loader. It was the height of impropriety to arrive with your own staff unless such arrangements had been made. It was also polite to arrive at the pre-arranged time for, as *Complete Etiquette for Ladies and Gentlemen* advised, 'Nothing is more annoying for the host who occupies a remote house than to have to make several long and fruitless journeys to pick up a guest at the station.'

On arrival at the house, whether by train or by motor car, a manservant took the guests' boxes up to their rooms while maids transported ladies' wraps and items such as books and papers. If a lady had not brought her own maid with her, the maid of the hostess took a guest's dressing case up with her, including the keys to any luggage, but never those of her jewellery case. It was then the maid's duty to show the visitors the location of the bathroom and where any items removed from their boxes had been placed – unpacking being a servants' duty, not that of the guest.

During the visit the lady's maid brought tea or chocolate to a female guest as requested and turned on the bath at a specified time. If the guest wished to ride the maid put out her riding habit, hat and other accoutrements and if necessary helped her on with her boots. Any clothes needing attention during the day were removed for brushing. In the evening she prepared the bedroom for the night, making sure that hot water was available, a fire lit and night attire laid out on the bed.

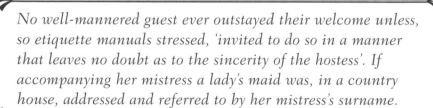

No well-mannered guest ever outstayed their welcome unless, so etiquette manuals stressed, 'invited to do so in a manner that leaves no doubt as to the sincerity of the hostess'. If accompanying her mistress a lady's maid was, in a country house, addressed and referred to by her mistress's surname.

A ROYAL VISIT DEMANDS ETIQUETTE OF THE HIGHEST ORDER

Every moment of such a country house visit, as well as the planning, demanded extreme attention, to avoid offence to either the members of the royal family or the staff travelling with them.

From the Victorian era, royal guests were most likely to arrive at their venue by train. Local volunteers might form a guard of honour to greet them but, said *The Lady's Realm* of 1904, detailing correct behaviour for a visit from Edward VII and Queen Alexandra, 'no band must be provided or procession arranged'. The King and Queen would be taken to their venue in a carriage drawn by four horses, with additional carriages being provided for royal staff and for luggage. On arrival, says the piece: 'The hostess will meet her Royal guests at the entrance to the castle or mansion in the usual way.'

> *To be suitably dressed in every way, and to cope with every eventuality, house guests were advised to take with them both mourning and half mourning to fit, if necessary, with the exact dress of the Monarch.*

Next day, breakfast was served to the guests in their own apartments and, unless it was the shooting season, they might stay in their rooms for most of the morning. Luncheon would be taken with any other guests. When, in the Edwardian era, many small tables were laid for this meal, host and hostess sat with their Majesties, 'with others of the company being honoured by invitations in turn'. Whatever the activities of the afternoon – anything from strolling in the grounds to the planting of memorial trees – the whole party would meet for an informal afternoon tea, which was 'always the occasion for a pleasant exchange of courtesies and cheerful conversation'.

Should a house party be involved – but only if this was the stated wish of the royal guests – it was left to the host and hostess to draw up the invitation list, relying, as *The Lady's Realm* of 1904 says, 'on their good taste in the matter', with the caveat that they '… may not be conversant with some little matter which may have come to the ears of the King's entourage, and so an undesirable may occasionally be included'.

During the day, royal visitors had the freedom to visit any part of the house or garden as they wished. When Princess Victoria stayed at Chatsworth in 1832, aged 13, for her first dinner party, she inspected the kitchen and pronounced it 'superb' both in dimensions and cleanliness. To keep the gardens perfect for her visit, head gardener Joseph Paxton employed the services of around a hundred men to remove any fallen leaves and branches and to keep the paths clear and the lawns perfectly rolled.

∽ A ROYAL DINNER ∾

The timetable for a stately country house dinner with King Edward VII and Queen Alexandra:

8.55 All guests assemble in the saloon or drawing room and form up in an avenue with ladies on one side and gentlemen on the other.

9.00 King and Queen arrive. His Majesty leads in the hostess, the host the Queen. They walk through the avenue of bowing guests, who then follow in order of precedence, with gentlemen offering their right arms to ladies.

9.15 Dinner is served with the King and Queen being waited on by their own servants, receiving dishes brought to them by servants of the household.

10.00 or thereabouts: The King exits the dining room. The ladies go to the drawing room. Gentlemen follow the King to the smoking rooms.

10.30 The King and gentlemen join the ladies, at which point additional guests may have arrived to be presented to the royal couple. Entertainment may then follow, such as private theatricals.

TIPPING IS A SERIOUS ITEM IN THE EXPENSES OF COUNTRY HOUSE VISITS

House staff might well expect to receive tips from guests, but it was always wise to check with the host, who might disapprove or expressly forbid it.

The practice of tipping began in coffeehouses and taverns in 18th-century England, and signs 'to insure promptitude' were prominently displayed to encourage patrons to tip and so hasten service. In the same period, it also became accepted that country house guests would tip the servants of their hosts; the money given to the servants was known as 'vails'. It was customary, when a guest was leaving the house, for the servants to line up in a double rank outside the door and for the guest to proffer a vail to each in turn. This could certainly be expensive. A German nobleman, Baron de Pollnitz, having visited an English stately home, stated bitterly that 'If a Duke gives me Dinner four Times a Week, his Footmen would pocket as much of my Money as would serve my Expenses at the Tavern for a Week.'

During the 19th century the tipping of servants became more random, but in the Edwardian era tipping once again became formalized and also more generous, the amount given depending, so Madge (Mrs Humphry) wrote in *Every Woman's Encyclopedia* in 1910, 'on the circumstances and particularly on the position and social standing of the visitor'. She also described a practice carried out in a few country houses: 'On the day when a guest terminates a visit the menservants are allowed to throw themselves in his or her way and they have to be tipped.'

In houses where tipping was forbidden the good mistress recompensed her staff with special monetary arrangements depending on the number and status of the guests entertained. If she failed to do this, staff would feel ill used indeed.

∽ How much to tip ∽

'Madge' advised these amounts for tipping in 1910. In addition tips would be given to servants who cleaned boots. Tips for bedroom servants would be left in the bedroom.

Butler: a sovereign for a few days' visit.

Chauffeur: from half a sovereign upwards if there have been many motor car rides, but if he only meets guests at the station and returns them there, five shillings or three half crowns.

Maid: five shillings to the maid who looks after a woman visitor's room.

Footman: half a crown for carrying luggage.

Parlour maid: also half a crown for luggage carrying.

THERE ARE MANY TALKERS BUT FEW WHO KNOW HOW TO CONVERSE AGREEABLY

Skill in the art of conversation was a measure of good breeding and an obligatory accomplishment for anyone visiting a country house. Politics and religion were always subjects to be avoided.

For both men and women the advice was to speak distinctly, and neither too fast nor too slowly. You should, said *Enquire Within*, 'accommodate the pitch of your voice to the hearing of the person with whom you are conversing,' and 'never speak with your mouth full'. As for jokes, it said that you should 'laugh after telling them, not before or during their relation'.

Women wishing their conversation to be agreeable were given strict instructions to 'avoid conceit or affectation, and laughter which is not natural and spontaneous ... Her lips,' the guide adds, 'will readily yield to a pleasant smile; she will not love to hear herself talk; her tones will bear the impress of sincerity, and her eye kindle with animation as she speaks.' And there was a strict warning against interrupting, considered to be extremely rude, and of 'pushing, to its full extent, a discussion which has become unpleasant'.

For men, it was unacceptable to show off with quotes in Greek or Latin or to indulge in pedantry. Men were counselled: 'If you feel intellectual superiority to any one with whom you are conversing, do not seek to bear him down; it would be an inglorious triumph, and a breach of good manners.' It was always unacceptable for house party guests to form themselves into 'sets', such behaviour being 'an affront to their host'.

Exceptions to the rules might apply to young men of high birth, as Society Small Talk *of 1879 explained: 'The broad and airy compliments of which men of a certain standing consider themselves privileged to pay to young ladies are not to be paid indiscriminately by ordinary mortals, the right of doing so belonging to those men who have a recognised reputation for this style of persiflage [banter].' For both sexes a ball was an occasion on which small talk was totally acceptable and conversation 'not expected to soar above the polite'.*

THE HOSTESS IS AN OSTENSIBLE CHAPERONE AT PARTIES TO WHICH GIRLS ARE ASKED WITH THEIR MOTHERS

Meaning that if the mother was otherwise engaged, it was the hostess's duty to make sure that a girl was properly chaperoned and protected from the undue attentions of men regarded as unsuitable.

On country house visits, there were various occasions on which young women could escape from close scrutiny, for even the best chaperone could not supervise her charge twenty-four hours a day. Outdoor activities such as tennis and hunting provided good opportunities for mixing with men whom chaperones might consider unsuitable – roués, upstarts, married men and those deemed to have limited prospects in the marriage market.

For the extremely daring there was scope at night for 'bedroom hopping', which might even be aided by the hostess, as Vita Sackville-West described in her novel *The Edwardians*: 'The name of each guest would be neatly written on a card slipped into a tiny brass frame on the bedroom door. This question of the disposition of bedrooms always gave the duchess and her fellow-hostesses cause for anxious thought. It was so necessary to be tactful, and at the same time discreet. The professional Lothario would be furious if he found himself in a room surrounded by ladies who were all accompanied by their husbands.'

Chaperones or not, country house parties were instrumental in the marriage market, coming as they did after the London Season (or for the less wealthy and well connected a series of summer balls and dinners) at which great efforts were made for eligible young men and women to meet. Hostesses with an eye for matchmaking would draw up their guest lists with this in view, particularly if they were mothers to sons and heirs who had not yet found suitable brides.

Snobbery came into the equation too, some ambitious mothers regarding any man but the heir to an estate as totally unacceptable for their daughters. And many young girls brought up in the country were likely to have a limited experience of life and little self-confidence, making them easily dazzled by men with polish and charm.

WHEN LONG GLOVES ARE WORN THEY SHOULD BE KEPT ON UNTIL ALL THE GUESTS ARE SEATED, WHEN THEY MAY BE REMOVED

Long gloves were essential women's evening wear for a formal dinner party, and the timing of their removal before eating and drinking was a matter of strict etiquette.

With full-length evening dress, long gloves were *de rigueur* and, whether they had buttons or not, were traditionally measured in terms of 'buttons' (one button is approximately an inch). An elbow length glove was a 16-button, a

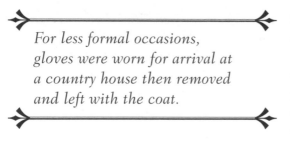

For less formal occasions, gloves were worn for arrival at a country house then removed and left with the coat.

mid-biceps glove a 22-button and a shoulder-length glove a 30-button. Good manners demanded that the only jewellery worn over a glove of any length should be bracelets – certainly not rings. When removing gloves they needed to be carefully folded down to the wrist before being delicately pulled from the tips of the fingers and thumb to avoid turning them inside out. On no account should the teeth ever be used to remove them. Finally they were expected to be laid across the lap, under the table napkin. After dinner, ladies would replace their gloves for dancing but keep them off for playing cards.

COURSES AT DINNER SHOULD BE CONSUMED IN THE CORRECT ORDER

The rule for dining in an age when food was served à la Française, but also when guests were partaking of a buffet laid out for a ball or wedding breakfast.

Until the style of dining changed to service *à la Russe* in the 19th century, food

Polite guests would serve themselves from the dishes placed nearest to them on the table and, if necessary, pass dishes to their neighbours and help them with carving and cutting as necessary.

for a formal dinner, including desserts, was placed on the table at the start of the meal, along with candles and other ornaments, plus flowers, pyramids of fruit and salt in fancy containers known as 'salts'. For perfect style, the table was set out symmetrically with many dishes placed in pairs. Dinner then took a set pattern. First to be served was the soup, in two or more varieties, ladled from large tureens. Following these were the 'removes': dishes such as fish, cutlets and tongue, plus roast turkey, mutton or chicken. Eaten with these were *hors d'oeuvres* (literally outside the main dish) placed around the table, which might include dishes such as small pies, oysters, eggs, and radishes. When all these had been cleared away, the

∽ DINNER IN VERSE ∾

Amazed at the array of dishes set before him at a dinner party held by Lord and Lady Amundeville in his satiric poem 'Don Juan', Byron penned the lines:

Their table was a board to tempt even ghosts
To pass the Styx for more substantial feasts.
I will not dwell upon ragoûts or roasts,
Albeit all human history attests
That happiness for man – the hungry sinner! –
Since Eve ate apples, much depends on dinner.

second course or *entremets* was served, consisting of large roasts accompanied by both vegetables and sweet dishes – creams, jellies and ices. The tablecloth was then removed; it was either replaced with a clean one or the final course, the dessert – of cheese, pastries, sweet concoctions and fruit – was placed directly on the table.

For a ball supper or wedding breakfast, the food was laid on a table or sideboard, with large joints carved on the sideboard or in the kitchen ahead of serving. Mrs Beeton in her 1861 bill of fare for a winter ball supper for 60, included more than 30 different dishes including: lobster salad; prawns; tongue, ornamented; boiled fowls with béchamel sauce; boar's head garnished with aspic jelly; mayonnaise of fowl; galantine of veal; roast pheasant; small ham, garnished; larded capon; raised game pie; Swiss cream; meringues; custards in glasses; fruited jelly; vanilla cream; biscuits and raspberry cream.

FOOD IS SERVED FROM THE LEFT, WINE FROM THE RIGHT

The unbreakable rule for service at table when each course was served separately. There would always be flowers on the table, along with candelabra and other ornaments.

Compared with dining *à la Française*, the 'new' *à la Russe* style of service demanded many servants to be on hand to bring in the dishes for each course. The rule was that food would either be offered from a large dish from which diners could help themselves or be served by the footman. Oysters, however, were an exception since, as *The Servants' Practical Guide* of 1880 says: 'To hand a large dish of oysters reposing on a serviette to each guest in succession is worse than bad style; it is awkward and inconvenient for the guests to help themselves from the dish.' Instead, they were placed in position before guests entered the dining room.

'In the case of there being two or more footmen or men-servants in attendance,' said the *Guide*, ladies seated on the right and left hand of the host were served first, followed by each guest in succession. 'It is not the practice,' it advised, 'to serve ladies before gentlemen when there are a number of guests present; the doing so would occasion no little confusion and loss of time.' If there was a choice of dishes, the footman asked the diner which they preferred. So, for the soup, 'Mock-turtle or Palestine, ma'am?'

Serving drinks was the butler's duty, with guests offered choices as appropriate. At a Victorian dinner, hock or Chablis was served with oysters, sherry with soup, and champagne, opened on the sideboard, with the fish and meat. Claret would also be offered 'throughout dinner in lieu of any other wine'. While dessert was being handed to guests the butler should follow, says *The Servants' Practical Guide*, 'with the claret and sherry, of which he offers the choice. When he has made the round of the table, he places two full decanters of sherry and a claret jug of claret before the master of the house, and thus, having completed the duty of waiting at table, he leaves the dining-room, followed by the other servants.'

Noiseless service, apart from necessary questions to diners, was obligatory for servants. No noise or clatter of any kind was acceptable. Servants were expected to move quietly, briskly and silently around the table.

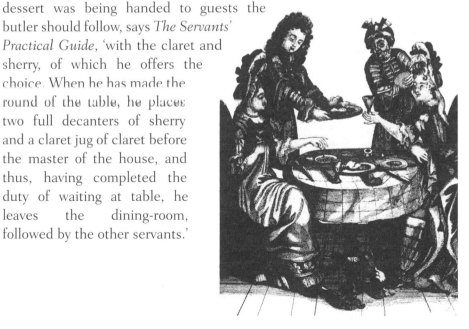

FOLLOWING DINNER, THE LADIES SHOULD WITHDRAW TO THE DRAWING ROOM

Here they were served tea, coffee and possibly liqueurs, leaving the men to enjoy port, brandy and cigars, along with fruit and nuts, either in the dining room or a smoking room.

It was essential to read the signs and to know when to make a move, since the signal to leave might be a subtle nod or smile to the senior lady guest – the one who had been escorted to dinner by the host. The hostess was instructed to time the exit carefully, avoiding the moment when a guest 'has just laid down a knife or fork or a wine glass, lest it might appear that the whole party had been waiting for the conclusion of that one individual's meal'. Also, 'if any one is in the midst of an animated or interesting conversation, the move must be deferred until it slackens off a little.' If, however, the mood of the dinner was

> *To concur with the rules of etiquette, ladies left the dining room in the same order as they had entered it, according to rank and age, and with married women – including widows – taking precedence over the single.*

in danger of being marred by any 'disagreeable or unwelcome topic' then 'withdrawing from the table promptly might turn out to be a great advantage'.

Conversation among the ladies would undoubtedly have been light and even gossipy. This might not be to everyone's taste, as Jane Austen, complaining of the 'poverty of conversation', noted in *Sense and Sensibility*: 'The gentlemen had supplied the discourse with some variety – the variety of politics, inclosing land and breaking horses – but then it was all over, and one subject only engaged the ladies till coffee came in, which was the comparative heights of Harry Dashwood and Lady Middleton's son William, who were nearly of the same age.'

IT IS ESSENTIAL THAT THE PORT IS PASSED CORRECTLY AND THAT CIGARS ARE WELL STORED

The perfect country house host still makes sure that everything served to his guests after dinner is of the best possible quality.

Vintage port was decanted by the butler before dinner to free it of both crust and sediment. First to take the port decanter is the host, who pours a glass for the guest on his right. He then passes the decanter to the left. Each person fills his own glass, then hands the decanter to the next person to the left. When the decanter reaches the host he pours himself a glass. As the decanter is passed

it should never touch the table or be lifted over a glass.

Cigars made of choice tobacco would be proffered to every male guest, and the host (and butler) would have made sure that they were stored in an airtight cedarwood box kept at an even temperature. The polite guest would trust his host to provide cigars of the highest quality. It was the height of bad manners for a guest to put his cigar to his ear and roll it in his fingers to test the dryness of the leaf.

Over port and cigars men would continue in political and possibly ribald conversation. But while in the early 19th century they might be summoned by a butler to join the ladies for coffee, by the 1920s they were expected to be polite enough to end their own gathering and adjourn in good time.

Whoever was serving at table – either butler or footman – needed to make sure that a cutter was to hand for clipping the 'cap' of the cigar and to offer a taper to light it. What was most important, as any well-bred gentleman would have known, was for the cigar to be evenly lit all round, so that it did not burn raggedly. Removing the band from the cigar was acceptable as long as this was done a few minutes after the cigar had been lit and it could be taken off without tearing the wrapper.

THE BUSINESS OF THE HOUSE SHOULD NEVER BE DISCUSSED WITH STRANGERS

Just one of many rules country house staff had to follow regarding conversation. Servants were also expected to be as unobtrusive as possible on every occasion.

Discretion was a quality of paramount importance in the country house, and it was obligatory that family matters, which staff might well overhear, were not

broadcast to visitors, including the servants of house guests, let alone to any visiting tradespeople or casual labourers. It was unacceptable for staff even to let it be known, by some facial expression, that they had heard the conversation of a family member or guest, or were aware of talk at the dinner table. Doubtless such matters were discussed in the kitchen or servants' hall, at least beyond the hearing of senior staff.

In 1901 the Ladies' Sanitary Association published *Rules for the Manners of Servants in Good Families*, many of which applied to the spoken word. Among them were:

'Nosiness is considered bad manners.

'Always move quietly about the house, and do not let your voice be heard by the family unless necessary. Never sing or whistle at your work where the family would be likely to hear you.

'*Do not* call out from one room to another; and if you are a housemaid, be careful not only to do your work quietly, but to keep out of sight as much as possible.

> *Failing to pay due respect to an employer could be a sacking offence, as was a failure to obey orders. Less serious, but unacceptable, were door banging and casting disrespectful looks.*

'Never begin to talk to the ladies or gentlemen, unless it be to deliver a message or ask a necessary question, and then do it in as few words as possible.

'*Do not* talk to your fellow servants, or to the children of the family in passages or sitting rooms, or in the presence of ladies and gentlemen, unless necessary, and then speak to them very quietly.

'Always speak of the children of the family as "Master", "Miss".'

Male servants were allowed to behave rather more assertively, although their position might make them seem servile. However, work in a country house was considered, for a country boy or young man, a worthwhile lifetime career, and most country house staff were much more contented than their city counterparts, not least because landowners made efforts to make them feel part of a worthwhile community.

THERE IS AN ORDER OF PRECEDENCE IN THE SERVANTS' HALL

Even below stairs, where the senior servants or 'pugs' set the rules for correct behaviour, there was a strict hierarchy among the staff, particularly at mealtimes.

In large country houses with many staff, the most senior members, known as the 'Upper Ten', were privileged to have their own dining room, disrespectfully referred to as the 'Pugs' Parlour', which might be the room of the housekeeper or the house steward. Often the 'pugs' dined here for an entire meal, being waited on by the steward's footman. If senior staff joined their junior colleagues for dinner, the ritual was that the 'pugs' entered the servants' hall for the first part of the meal, arriving in formal array and in order of precedence, while the lower staff stood in deference.

All then took their seats at the long table, with the housekeeper at the head. To her right would sit the cook, to her left the lady's maid. On each side of the table women staff sat by rank, creating an all-female congregation. The other end of the table would be headed by the steward or butler with the under butler to his right and the coachman to his left. As with women staff, the men sat by rank, in descending order, so that men and women – in reality most probably boys and girls – of lowest rank were placed in the middle.

To begin the meal, grace was said by the most senior person, then the meat course was served. Any carving needed was the butler's duty. The 'pugs' then retired to their own dining room, taking with them the necessary china and glass to consume dessert and cheese, washed down with wine. On their way out of the hall they might express their disdain at drinking the beer provided by tipping any that remained in their glasses into the sink – a ritual dubbed 'sinking the beer'.

CHAPTER 5
ENTERTAINMENT, LEISURE AND SPORT

*T*he country estate was, and remains, the perfect place for relaxation and sport, from hunting, shooting and fishing to tennis, croquet or a leisurely stroll around the grounds. Its facilities were used and appreciated by the family and their friends, but it was also ideal for entertaining those people the master and mistress of the house wished to court for social, political or commercial reasons, from royalty to the captains of industry. Indoor pursuits were also an essential part of a country house visit, and could include cards, charades and musical entertainment. And within the confines of a country house gambling games illegal in other settings could be played without compunction.

Many of the sporting activities of the country house provided opportunities for men and women to mix freely in a way that was impossible in more formal circumstances. The house party was an accepted way for eligible and 'suitable' bachelors to be introduced to girls considered to be in need of good husbands.

For the country house servants, leisure time was minimal and much appreciated. Since most of the junior staff were recruited locally, days off were an opportunity for them to visit their families or to make excursions to local shops and hostelries. In most houses it was also possible for visitors to be entertained in the servants' quarters, as long as no improprieties took place. Traditionally, Twelfth Night was the date for the annual servants' ball, with family and staff reversing roles for the night and mingling on the dance floor.

THERE IS AN ABIDING CHARM TO A GARDEN PARTY

Inviting guests to take tea in the garden became common country house practice in the Victorian era. Dress, except for those exerting themselves in sport, was decidedly formal.

Typical advice to garden party hostesses was that a garden party should be synchronized with a display of roses or some other favoured flower and, given the vagaries of the British climate, that indoor rooms should be made ready in case of rain. Guests would be invited to arrive at 3.30 or 4.00 and be greeted by their hostess in a shady spot, and tea would be served in bone china cups at small tables set with good linen. The hour for departure would be 7.00, after guests had taken tea, played games such as tennis and croquet if they so wished and, at a grand party, been entertained by a band playing popular music.

For the Victorian garden party, dress was as formal as for an indoor tea party, but daintier. Women would wear long dresses of silk or chiffon (pale colours were acceptable for all except those in mourning), plus hats and gloves. Parasols were carried to protect complexions from the sun, it being

As well as sandwiches and cakes, strawberries and cream – using home-grown fruit from the kitchen garden – jellies and ices were popular foods for summer country house garden parties, all served by the house staff under the direction of the butler. Making ices was possible thanks to the ice houses constructed in the grounds and kept filled all year round (see Chapter 6).

considered indelicate for a lady to allow her skin to tan. Men were expected to don suits or military uniforms, as appropriate. Similar rules pertained in the Edwardian era and it was only in the 1920s that garden party dress became less restricting, when fashion changed radically in favour of looser dresses, often bias cut and with chiffon overlays that created a floating, fluttering effect. Floral dresses became the vogue in the 1950s, when hats were still worn for formal garden parties.

GARDEN PARTY SANDWICHES

The sandwich, made using white bread and with the crusts removed, was the staple of the garden party, but until the mid-19th century the filling would have been ham, tongue or beef, since cucumber was considered both cold and poisonous or, according to the last verse of a poem published in Punch:

So much for cold John Cucumber,
Whom few insides can stand,
Of all the Cucurbitae
The worst in Merrie England.

Gentleman's Relish, or Patum Peperium, an anchovy paste invented in 1828 by John Osborn and made to a recipe still carefully guarded, was another popular sandwich filling.

CROQUET MAY BE PLAYED BY PERSONS OF ALL AGES AND OF EITHER SEX

So said the Victorian manual Enquire Within, *adding, 'it is especially adapted for young ladies and young persons, as it demands but slight personal exertion, while it affords delightful and health-giving sport'.*

By the 1860s croquet had become the number one outdoor social pastime, and every big house would have a set of balls, mallets and hoops to set out on the lawn. One of the earliest croquet lawns was at Cassiobury Park in Hertfordshire, owned by Arthur Algernon Capel, sixth Earl of Essex. The Earl, a leading entertainer in high society, caught the 'croquet bug' in the early 1860s and not only hosted lavish croquet parties but marketed the Cassiobury brand croquet set, manufactured in his own sawmills.

Unlike tennis, croquet could be played in ordinary outdoor dress – tweeds or a blazer and boater being suitable for men. It was thus perfect garden party entertainment and a chance for unmarried men and women to meet free from the close scrutiny of a chaperone.

FOUL PLAY

Among its many instructions for croquet players, Enquire Within of 1894 included the following in its list of 'foul strokes':

- To spoon, that is, to push the ball without an audible knock.

- To strike a ball twice in the same stroke.

- To stop a ball with the foot in taking a loose Croquet.

- To allow a ball to touch the mallet in rebounding from the turning peg.

- If a player, in striking at a ball which lies against a peg or wire, should move it from position by striking a peg or wire, the ball must be replaced, and the stroke taken again.

IT IS FASHIONABLE FOR THE ARISTOCRACY TO ADD TENNIS COURTS TO THEIR COUNTRY HOUSES

Or so it was in the Victorian and Edwardian eras, when lawn tennis became popular for both men and women, helped by the invention of the lawnmower in 1830, which allowed grass to be cut short.

Frequent rolling, which would be done by one of the groundsmen, was also necessary to keep the court smooth. Typical of its time was the tennis court created at Hatfield House in 1842 by the second Marquis of Salisbury, who played his first game there against the Rev F.G. Faithfull, then Rector of Hatfield. The lawn was marked out with white chalk, although in some places

In 1845, Queen Victoria was a spectator at a game of tennis at Stratfield Saye, home of the Duke of Wellington on the Hampshire/ Berkshire border. In her diary she recorded 'a very fine game played between Ld. Charles, the Duke's marker, & a fat man called Philips, the Duke's butler, who plays beautifully'.

tapes were pinned to the ground for the purpose. Now part of a private club, the court is still played on.

The good hostess was sure to have enough racquets for everyone and ensured that her guests were well informed about dress. Until 1910 women wore hats, tight-sleeved jackets and heavy skirts with tight waists, while men donned full-length white trousers and long sleeved shirts. Rubber-soled shoes, introduced in 1867, were *de rigueur* for serious players. More amateur performers wore boots or substantial shoes.

On court it was deemed impolite to make audible comments disparaging others' play or to lose one's temper. At the end of a match, shaking hands and thanking one's partner and opponents were also essential, followed by a substantial but informal tennis tea, as for a garden party, served outside on good china.

THERE ARE FEW MORE DELIGHTFUL ENTERTAINMENTS THAN A CRICKET MATCH

Although it began as a working man's game, cricket soon became popular in the perfect setting of the country house.

Some of the earliest country house cricket matches were held in the mid-18th century at Knole in Kent, home of the first Duke of Dorset, where the pitch was maintained in prime condition by Valentine Romney, one of the best players of the time. Some grounds had pavilions added, like the exceptional example at Hemingford Park in Cambridgeshire. Here, matches even took place against county sides, when gentlemen (i.e. amateur) players, including such great names as W.G. Grace and C.B. Fry, stayed as guests in the main house while any professionals were accommodated in one of the three bedrooms on the first floor of the pavilion.

A cricket match was a social occasion – often the focus of a country house weekend – and a chance for men and women to parade in their finery. Between innings, country house cricketers might be entertained to a large lunch. Lord William Lennox, writing in the 1870s, sets the scene: 'A party sallied forth from the house headed by my father …

Gentlemen often played for high stakes. Lord Frederick Beauclerk, a clergyman who eventually became president of MCC in 1826, estimated his income from side bets at cricket matches at £600 a year.

They had quitted the dining-room after imbibing a fair quantity of port wine, leaving instructions to the butler that clean glasses, devilled biscuits, and a magnum of "beeswing" [a well-aged port complete with crust] should be ready on their return from the cricket field.'

FISHING IS AN EXCELLENT SPORT FOR GUESTS

Around the country house, fishing could take place from a lake in the grounds, or from a river or stream. Fishing from a boat was particularly pleasurable.

At the grandest houses, freshwater fishing expeditions began and ended at a pavilion or room on the lakeside, from which boats could be launched and where the catch could be cooked and eaten immediately. This kind of fishing was an acceptable entertainment for women as well as men. Kedleston Hall in Derbyshire boasted a

⬷ IN PRAISE OF FISHING ⬷

At Antony House in Cornwall in 1610 the translator and antiquary Richard Carew composed this poem in praise of his fishpond, which also tells of the species it contained:

My fishful pond is my delight
There sucking mullet, swallowing bass,
Side-walking crabs, wry-mouthed fluke,
And slip-first eel, as evenings pass,
For safe bait at due place do look,
Bold to approach, quick to espy,
Greedy to catch, ready to fly.

magnificent fishing pavilion designed by Robert Adam, ornamented with stone carvings, still-life paintings of fish and seascapes. There was a Venetian window on the upper storey from which women could fish in comfort away from the sun (lest their complexions suffer). Below, at water level, were two boathouses and a plunge bath.

The good country house host kept his waters well stocked with fish and supplied rods for guests – and possibly fishing gear such as waders as well, since fly fishing for salmon and trout often involved wading in chest high. In the winter months, anglers rubbed their legs with goose fat to protect themselves from the cold. Whisky was kept to hand to warm fishermen as they retreated regularly from midstream.

BILLIARDS MAY BE PLAYED BEFORE OR AFTER DINNER

Popular all year round as a country house entertainment, early evening billiards would always be halted by the bell warning guests to retire to their rooms to change for dinner.

Billiards began as a game for both sexes, but in houses where the billiard room doubled as a smoking room it was a more male preserve. As well as following the basic rule of stooping so as to get

Among families, a fast and furious game called billiard fives developed, as described by Deborah, Duchess of Devonshire, in her memoirs. In this, the prime objective was to hit the ball down the table fast enough so that as it returned, the person running behind you had no hope of hitting it, thereby losing a point.

eyes, ball and object ball in a straight line, it was essential to take care never to commit the cardinal sin of mis-cueing and ripping the green baize (felt) covering the table. It was also bad manners to stand close to a player, or to make a noise, such as by lighting a match, while he was making a shot.

In some country houses the hall served as a billiard room, being the only space large enough to accommodate the huge table with its massive legs. But from the 18th century, as the game became increasingly popular, it was often afforded its own space, ideally lit from above.

 CARE OF THE BILLIARD ROOM

In his 1884 guide to the game William Cook proffered sound advice to players and the servant in charge of the room:

- *Cues should always be kept in an upright position and replaced in the rack immediately after play.*

- *Always brush the cloth every time the table has been played on, brushing with the nap and pushing dirt into the pockets.*

- *Iron the cloth regularly.*

- *Avoid letting the room get too cold. If necessary warm the cushions with tin boxes filled with hot water designed specifically for this purpose.*

A SHOOTING PARTY REQUIRES A SUSTAINING LUNCHEON

And more besides. A day's shooting was hard work for country house staff needing to ensure an excellent day for all concerned.

In the kitchen, a shooting day might begin as early as 4.30am, when staff would prepare sandwiches for the beaters and begin making elaborate luncheon for the 'guns', their lady companions and the gun loaders. The food needed to be transported, along with trestle tables and chairs, tablecloths and other linens, plus glasses and cutlery, to a suitable outdoor spot. If the weather was inclement, arrangements were made for the meal to be served in the gamekeeper's cottage.

Since a shoot normally lasted three days (traditionally Tuesday to Thursday) variety was also essential. Popular hot luncheon dishes included stews, casseroles, curries, suet puddings and roast meats accompanied by baked potatoes; typical cold fare would be game pie, ham and other meats.

Apple dumplings or turnovers, rice pudding and rib-sticking steamed puddings such as spotted dick made ideal desserts on winter days. To drink, the butler and footmen – who always served the meal – would proffer wine, beer or cider. While guests were enjoying their feast, the loaders dined on meat and jacket potatoes while the beaters had bread and cheese.

A pony and gig was needed to ferry the luncheon to the shooting party. Hot food was piled into sacks or put into pans, which were then inserted into wooden 'hot boxes' insulated with padding.

While previous generations shot largely to keep down vermin and to kill game, shooting on country house estates became high sport after the Prince of Wales (the future Edward VII) purchased the Sandringham estate in Norfolk in 1862. Here literally thousands of birds, rabbits and hares were shot each year according to a strict calendar. After each day's shooting (see also Chapter 4) staff were required to count and hang birds in game larders and pluck any that were to be eaten for dinner that night, before cleaning muddy boots and brushing down clothes.

HUNTING IS BOTH A SPORT AND A SOCIAL OCCASION

By Edwardian times both men and women were able to participate more or less equally in the field, if not in the celebratory aspects of the day.

The first day of November was the start of the fox-hunting season. The men of the household would eat a cooked breakfast before venturing out into the field. Ahead of the hunt, vintage port was served indoors to male members of the family and any hunt dignitaries, while outside the house the butler and footmen served a lesser variety of port to the men. Later in the morning the whole party returned for a hunt breakfast 'proper' in the dining room, to be served – again to the men only – by the butler and to include wine, beer and spirits. The women who had been hunting breakfasted more soberly in the drawing room with the mistress of the house.

Riding to hounds demanded particular skill from women, who were obliged to ride sidesaddle, dressed in long skirts. But because they were not chaperoned, they had the opportunity to talk openly with their male companions.

Women took great pleasure in the freedom of riding in the countryside and leaping hedges, gates and brooks. But it could be dangerous, as Anthony Trollope describes in his 1872 novel *The Eustace Diamonds*: 'To Lizzie it seemed as though the river were the blackest, and the deepest, and the broadest that ever ran. For a moment her heart quailed – but it was but for a moment. She shut her eyes, and gave the little horse his head. For a moment she thought she was in the water … But she was light and the beast made good his footing and she knew she had done it.'

Later in the day celebration was paramount. Lord Hervey, guest of the 18th-century writer and politician Horace Walpole, described the evening celebrations as 'noisy, jolly, drunk, comical and pure merry ... We used to sit down to dinner a little snug party of about thirty odd, up to the chin in beef, venison, geese, turkeys, etc; and generally over the chin in claret, strong beer and punch.'

AFTER DINNER, CHARADES MAY BE ACTED IN COSTUMES

Charades came to the country house from the courts of Europe and were originally riddles, acted out as dramas, to be solved by the assembled company.

The charades of 19th-century Britain were elaborate affairs, and family members would not only rehearse but dress up in costumes to mime the word or phrase that had to be guessed. Plenty of props would be used, too. Other types of charades took the form of conundrums. In every case, each syllable of the word to be guessed was designated 'my first', 'my second' and so on. Some, like this one, with its virtuous somewhat melodramatic theme, were written in verse:

Contemporary books such as Household Amusements and Family Recreation, *published by Samuel Beeton, husband of the famous cookery writer, supplied texts, sometimes amounting to short plays, for actors to learn or recite.*

> *The birds, whose glad voices are ever a music delightful to hear,*
> *Seem to welcome the joy of the morning, as the hour of the bridal draws near.*
> *What is that which now steals on* my first, *like a sound from the dreamland of love,*

And seems wand'ring the valleys among, that they may the nuptials
approve?

'Tis a sound which my second explains, and it comes from a secret
abode,

And it merrily trills as the villagers throng to greet the fair bride on her
road.

How meek is her dress, how befitting a bride so beautiful, spotless and
pure!

When she weareth my second, oh long may it be ere her heart shall a
sorrow endure.

See the glittering gem that shines forth from her hair – 'tis my whole,
which a good father gave;

'Twas worn by her mother with honour before – but she sleeps in peace in
her grave.

'Twas her earnest request, as she bade them adieu, that when her dear
daughter the altar drew near,

She should wear the same gem that her mother had worn when she as a
bride of promise stood there.

And the answer … Ear-ring.

FOR LARGE PARTIES, THERE SHOULD BE A TABLE FOR CARDS, AND TWO PACKS OF CARDS PLACED UPON EACH TABLE

Playing cards was a popular after-dinner activity at the country house, with gambling allowed in some establishments.

A whole variety of card games were played, of which whist, a game for four people, was the most popular in the 1880s. Also enjoyed were loo, best with five or seven players, and vingt-et-un for two or more. Bézique, played with two packs of cards with the twos, threes, fours, fives and sixes removed, was another absorbing after-dinner game.

In the same decade, baccarat, a French gambling game played with cards on a specially designed table, was almost obligatory for the fashionable country house party hostess. The Prince of Wales was a particular fan of this illegal game, and until an incident at Tranby Croft in 1890 – where Sir William Gordon-Cumming, a fellow guest, was accused of cheating and subsequently attempted to sue the witnesses to his misdeed, creating a royal scandal – the Prince always carried his own set of chips ready for an impromptu hand.

Contemporary guide books set out the rules for playing whist and its etiquette. Players and onlookers were advised to keep silent, never to remark on a hand that had been dealt and to pay close attention to the game. Onlookers might, however, be appealed to as referees in case of disputes.

As a result of the Tranby Croft affair, the playing of baccarat was discontinued in society, but almost simultaneously bridge was introduced and became an overnight craze with women in particular. The game was so popular that *The Lady's Realm* even ran an article entitled 'Is Bridge Immoral?' in which, stating his case, one Adrian Ross writes: 'A guileless girl, staying at a country house of intimate friends, it appears, was asked to play bridge; she declined, not knowing the game, but was promised advice and yielded. After sundry rubbers, she went to bed; but was informed next morning that she had lost an amount variously stated as from sixty to a hundred pounds.' The outcome was that her father was called for, the debt paid, and the girl 'removed from the haunt of vice'.

A MUSICAL EVENING WILL ADD GREATLY TO ANY HOUSE PARTY

The full-blown musical evening was a formal affair that included supper, as for a ball, but music might also be played more casually following dinner.

Such was his enthusiasm for music that in the 18th century the Duke of Chandos, friend of the composer Handel, lodged an entire 27-piece orchestra at his house at Canons Park, north-west of London.

In a large house, a musical evening might involve performance in one room while noisy partying carried on elsewhere. Advising hostesses on such a situation, *Cassell's Household Guide* said: 'Where the necessity of receiving large numbers is met by adequately numerous apartments, suites of rooms, galleries and so forth, all conversation is not restrained when music commences; only those within earshot of the performers are compelled by courtesy to keep silence. In the adjoining rooms conversation has full sway, and although a distant murmur may penetrate to the music-room, listeners are not supposed to be disturbed by it.'

For a less formal affair, those guests who were interested in music would simply find places for themselves near the piano or other instrument. Conversation was permitted but good taste dictated that 'loud remarks and laughter are never more ill-timed than during the performance of music'. In

houses with galleries above the dining room, music might be played while guests were dining.

Well-educated girls of the family invariably learned the piano and were expected to play – and sing. Should any lady guest sing or play, and be asked to do so by her hostess, the gentleman nearest to her was expected to escort her to the piano, help her to arrange her music and make sure that her gloves, fan and handkerchief were taken care of. For the hesitant, *Enquire Within* advised: 'If you sing well, make no previous excuses: if indifferently, do not hesitate when you are asked, for few people are judges of singing, but everyone is sensible of the desire to please.'

∽ BEST BEHAVIOUR ∾

Some rules for good conduct at a musical evening:

* *Guests are greeted by the hostess, remaining in position until all have arrived.*

* *The rustling of programmes and tapping of fans are the only forms of applause permitted to ladies. On no account should they say 'bravo'.*

* *Sacred music is not appropriate at evening parties. If any is included it should be during the first part of the evening. Chamber music is best suited to a drawing room.*

* *The question of whether guests will perform at a musical evening should be settled by the hostess before the event.*

* *It is a compliment to professional performers to ask them to remain for any amusements that may follow.*

EMBROIDERY IS A SKILL AND TASTE MUCH TO BE DESIRED

A leisure activity for country house ladies above stairs, with practical uses such as decorating personal items of clothing. Below stairs, needlework was a duty for the upper housemaid (under the housekeeper's direction), the lady's maid and the housekeeper herself.

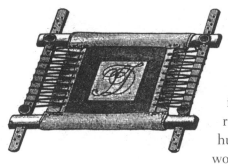

'Light or fancy needlework,' said Mrs Beeton, often formed 'a portion of evening recreation for the ladies of the household.' They might trim and decorate garments, adding their personal monograms, make useful items such as mats, antimacassars and table runners, or create decorative samplers to be hung in bedrooms. To make it easier to see her work by candle or lamplight, an embroiderer often had a large mirror on a stand placed strategically behind her to reflect light on to her work.

By Victorian times, ready-printed designs for embroideries were available, a trend denigrated by *Cassell's Household Guide*, which insisted that 'this absence of invention and good taste' was 'neither necessary nor desirable'. Advocating a return to 'the beautiful occupation of their female ancestors', the manual expressed the view that 'There is at the present time much desire for this shown among the upper classes, and legitimate embroidery is again rapidly becoming a fashionable employment.'

Ladies' maids spent much time altering and embellishing garments to keep up with changing styles and fashions, and even trimming hats. In large households, sewing of all kinds might be put out to women in a nearby village, or taken on by a maid hired – often from France or Switzerland – specifically for the role.

A COUNTRY HOUSE BALL IS A WELCOME DIVERSION

Once the nobility and gentry were able to travel with reasonable ease and began to socialize, private country house balls became regular entertainments, with guests ranging from local friends to royalty.

The earliest balls took on a distinct pattern, beginning with dinner, which might be accompanied by music. Guests then retired to the drawing room for dessert, to drink tea and possibly play cards before returning to the dining room for dancing. And to ensure that guests did not return home hungry, yet more food was proffered at the end of the evening.

By the 18th century, the most popular dances included the hornpipe, various reels and the minuet, a formal and difficult dance described by Edward Austen-Leigh, nephew of Jane Austen: 'The stately minuet reigned supreme; and every regular ball commenced with it. It was a slow and solemn movement, expressive of grace and dignity, rather than of merriment. It abounded in formal bows and curtsies, with measured paces, forwards, backwards and sideways, and many complicated gyrations.'

To assist coachmen bringing guests to and from balls, dates were deliberately arranged to coincide with the full moon to help them find their way more easily along country lanes.

As the century progressed, balls became ever more elaborate, with masquerades – to which guests wore masks – and *ridottos*, which were combinations of dances and concerts. At ball-assemblies dancing was combined with activities such as card playing, which carried on, with an all-evening supper, in different rooms of the house. By the Victorian era the grand supper ball was the height of fashion, with dancing in the ballroom, if there was one, and with dances including the quadrille, which according to etiquette (see Chapter 4) was danced by the lady of the house to open proceedings.

Good music was indispensable to the success of the ball. 'The pianoforte,' said *Cassell's Household Guide*, 'is not sufficient.' It also commented on the serving of food: 'It is imperative that any viands that have been cut should be replaced by fresh ones for succeeding visitors,' as to leave some of the company 'to fare as best as they can from remnants would be the height of ill manners.'

THE SIDEBOARD ARRANGEMENTS MUST BE OF AN EXTENSIVE NATURE FOR THE SMOOTH WORKING OF A BALL SUPPER

Advice for running a ball from The Servants' Practical Guide *of 1880, which spelled out arrangements down to the last detail.*

'It is the butler's duty to announce supper,' says the *Guide*, adding that 'he does not make the announcement as when announcing dinner, but informs his master and mistress that supper is served,' leaving them to inform their guests. Guests would then seat themselves at small tables and be served

champagne, opened by the butler and poured by his support staff. If, however, a servant was asked to leave an entire bottle at a table, 'that [a guest] might help himself and his friends at the same table' then this wish would be fulfilled without delay.

Food, which might have been prepared under the supervision of a chef brought in specifically for a very grand occasion, was served from the sideboard. Each person was either brought a small dish of food they had requested or helped themselves from a large dish. It was the fashion in the late 19th century to begin a ball supper with a choice of two hot soups served in soup cups, handed to guests 'in the order in which they are seated'. Following the soup, a hot entrée might be served, but a selection of cold meats – viands – was more usual, and might include boned turkey with truffles and a boar's head with pistachio nuts, as well as dishes such as lobster salad and plovers' eggs in aspic.

∽ On the Sideboard ∾

For serving a ball supper it was essential to have on the sideboard in addition to food:

A sideboard cloth, specially fashioned for the purpose.
Ornamental plates, china.
Large and small silver forks; large table knives; table and dessert spoons.
Champagne and wine glasses, tumblers.
Decanters of sherry.
Champagne in pails of ice.
Napkins ready to place on tables as required.

Vegetables were not served unless they formed or accompanied an entrée. To assist guests in their choice of food, a menu card might be placed on each small table seating four or six.

In the country a ball meant late hours. Supper was not served until 1.00 in the morning and lasted for about two and a half hours. After 3.00 there might be little food remaining, but champagne would still be freely provided until the last guest had departed, after which 'the eatables' were taken to the kitchen, fruit placed in the housekeeper's room and glass and plate removed to the pantry. The wine, always the responsibility of the butler, was put away by him.

A SERVANT'S BALL SHOULD BE HELD AT LEAST ONCE A YEAR

In large households, a ball or grand dinner for servants and tenants on the estate would be held at least once a year, usually after Christmas at Twelfth Night. Less formal dances were also regular entertainments.

To display their generosity, good masters and mistresses, assisted by other members of the family, even turned the tables and waited on their servants, as well as providing bountiful fare. At Hatfield House in the 1890s the Cecil family provided, each New Year, a 'tenantry dinner' consisting, so the records note, of 'two hundred and seventy-six meat dishes, forty-eight pie dishes and salts'. Three dozen butter boats and five hundred plates were needed for the occasion, which was marked with a speech by the landlord.

At Tredegar House near Newport the annual servants' balls, held on Twelfth Night, were famous in the area and were attended by estate workers and local suppliers as well as household staff. If he was in residence Lord

Tredegar would have the first dance with the housekeeper before leaving to allow everyone to enjoy the evening. A band played on into the early hours, the party finishing at about 5.30 in the morning, although some staff were due to begin work only half an hour later. A huge buffet was laid out in the great kitchen and wine and beer were freely supplied. Not surprisingly many servants met their future spouses at the balls, when they had the opportunity to relax and dance resplendent in their best clothes.

PREPARATIONS FOR CHRISTMAS MUST BEGIN WELL IN ADVANCE

In order, to quote Mrs Beeton, 'to meet old Christmas with a happy face, a contented mind and a full larder'.

In the kitchen, Christmas preparations began with delicacies such as cakes, puddings and mincemeat being made in November. At Hatfield House in 1877 ingredients used for the latter were 28lb (12.7kg) each of raisins, currants and moist sugar; 12lb (5.5kg) of mixed peel; 18 lemons; 1lb (450g) of mixed spice, and a good many

As well as being exchanged between family members, gifts were given to servants. These were often practical, such as new uniforms or, for female staff, the fabric with which to sew them. Money was also given, as at Woburn where footmen were each presented with a monogrammed envelope containing a five pound note, which would have been the equivalent of a month's salary. Gifts might also be distributed to the poor in the community.

apples; four bottles each of brandy and sherry; and two bottles of rum. On the day, the dishes served were roast goose and/or a roast sirloin or rib of beef or possibly a stuffed turkey. Desserts, in addition to the flamed pudding and mince pies, might include orange jelly, lemon pudding, almond pudding and a 'dish of snow' made by mixing puréed apple with sugar and beaten egg whites and served decorated with myrtle leaves.

A Christmas tree was a must after Prince Albert put one in Windsor Castle in 1841, although it

CHRISTMAS GAMES

As well as eating and drinking, Christmas entailed playing games including cards, draughts and backgammon. Others were more riotous:

Snapdragon – Nuts and sweets were snatched from a bowl of hot brandy.

Bullet Pudding – A Georgian favourite described by Jane Austen's niece Fanny Knight: 'You must have a large pewter dish filled with flour which you must pile up into a sort of pudding with a peek at top. You must then lay a bullet at top and everybody cuts a slice of it, and the person that is cutting it when it falls must poke about with their noses and chins till they find it and then take it out with their mouths of which makes them strange figures all covered with flour but the worst is that you must not laugh for fear of the flour getting up your nose and mouth and choking you: You must not use your hands in taking the Bullet out.'

was not brought indoors, along with traditional evergreen decorations – holly, ivy and mistletoe – until Christmas Eve. Ahead of the season children of the house were encouraged to make their own decorations, such as small paper boxes containing sugar plums.

A DAY OFF IS AN OPPORTUNITY FOR A FAMILY REUNION

For servants fortunate enough to be allowed sufficient time off, especially if they were young, reconnecting with the family was vital.

As a rule, younger and more junior country house servants were hired from nearby farms, villages and towns, making family visits possible despite the fact that these might entail considerable distances walked on foot. In benign households the cook might ensure that a maid was given a basket of food such as preserves, cold meat and cakes to take with her. Hours off were also the chance for couples to walk and talk together, and to progress through the rituals of courtship away from the prying eyes of the 'big house'.

When leisure time was spent in the house many staff enjoyed playing musical instruments such as fiddles. In some establishments a piano was provided in the servants' hall. Spare time might also be spent in sewing and knitting, reading the penny novelettes of the day, and similar domestic pursuits. The local pub was also a place of

> *Even if days off during the year were sparse, Mothering Sunday, the fourth Sunday in Lent, was the one day of the year on which girls in service were always allowed home to visit their families.*

entertainment. Writing in 1759 the owner of an estate remarked that: 'In the village where I live we have five public-houses, two of which have Nine-pin grounds and shuffleboards ... our livery servants neglect their household and other services, to spend their time at those houses.' Cards and dancing were other attractions of public houses frequented by country house servants.

Outdoors, male staff might take part in cricket matches organized by the master of the house or in a village. The Duchess of Portland, knowledgeable regarding the value of fresh air and exercise, presented each of her footmen with a bicycle and a set of golf clubs. Indoors, staff might be encouraged to join in family theatricals and concerts.

IT IS POSSIBLE FOR FRIENDS TO BE ENTERTAINED BELOW STAIRS

Depending on the strictness of the regime, staff could receive calls from friends and relatives on days off and at less busy times during the day.

During such visits, staff would talk, drink tea – or ale if available – and play cards. Meals might also be taken. In his 18th-century diary John Baker recounts the visit of a 'Mr Walter the exciseman and his wife and daughter' who called one evening at about 7.00 to play cards and drink tea with the housekeeper until 11.00 in the evening.

In country houses with large staffs, firm friendships were formed, especially between those of similar ages such as maids, who often shared a bedroom. But social life was more difficult for 'upper' servants, who were discouraged from fraternizing with the lower orders. In houses that were remotely situated grooms and under gardeners might help out in the pantry in the evenings purely for the company. Senior staff had greater privileges. In the late 19th century at Goodwood House in Sussex, for instance, there was no question that the steward was allowed to entertain whoever he chose to invite to his room.

Casual and part-time workers would be frequent visitors while country house weekends were ideal opportunities for staff to meet their counterparts from other establishments.

Although relationships between members of staff were discouraged – in many houses, for example, women servants were allowed into the servants' hall only for meals and menservants were discouraged from entering the kitchen or laundry – they flourished nonetheless. Clandestine meetings, often in the dark of early morning, were frequently arranged, but by the Edwardian era supervised meetings might be allowed in country houses, and a lady's maid might be permitted to invite friends for tea in her sitting room.

CHAPTER 6
GARDEN AND GROUNDS

*D*efining the 'English Gentleman' in 1728, Daniel Defoe said that he should have 'venison perhaps in his park, sufficient for his own table at least, and rabbits in his own warren adjoining, pigeons from a dove-house in the yard, fish in his own ponds or in some small river adjoining, and within his own royalty and ... all the needful addenda to his kitchen, which a small dairy of four or five cows yields to him'. As this description implies, every country house owner also needed to be something of a farmer in order to provide all that his family needed, as well as giving employment to many local people.

The way in which the country house garden was arranged evolved with the centuries. So while the Tudor garden was embellished with formal knot gardens and the like, and the 17th-century house and park were separated from each other by a terrace, the designers of the 18th century removed the barriers between the two, opening up the landscape, making the whole look more natural and adding eye-catching features such as follies and temples.

While the flower garden provided a supply of blooms for decorating the house, in the kitchen garden planting was arranged by the head gardener in order to cultivate fruit and vegetables for year-round use and, with the help of glasshouses and heating arrangements, to grow the exotic items so prized by country house owners. The grounds also needed to include facilities for brewing and an ice house, as well as providing space for the stabling and care of horses, the safe keeping of carriages and, in later eras, the garaging of the earliest motor cars. Important members of staff such as gamekeepers were provided with their own residences on the estate, while grooms would be accommodated above the stable block.

THE HEAD GARDENER MUST UNDERSTAND THE ELEMENTS OF LANDSCAPING

A vital aspect of the country house garden, but particularly so in the age of the great landscape designer Lancelot 'Capability' Brown and his successor Humphrey Repton.

The landowner keen to improve the look of his grounds would, if he did not employ a specialist such as Brown, rely on his head gardener to come up with suggestions for laying out walks, devising picturesque views (known as 'prospects'), arranging features such as urns, obelisks and benches, and rigging up

It is no coincidence that the landscape movement coincided with the Acts of Enclosure passed between 1730 and 1820, which allowed landowners to formalize the separation of agricultural land and park.

fountains and other waterworks. A gardener advertising for a position in the *Bristol Gazette* in 1773 was careful to state that he understood 'laying out Ground Work, in the modern Taste, perfectly well …'

'Capability' Brown earned his nickname from his skill in exploring an estate to assess its 'capabilities' or potential. His aim was to achieve a beauty that blotted out nature's imperfections to create an air of serenity. Under Brown's supervision, at locations such as Croome Court in

Worcestershire, where he also advised on the creation of the house itself, hills were levelled, lakes dug and grottoes created. To shape the contours of a garden to give the desired effect, he planted huge numbers of trees, even transplanting mature specimens.

Many of the great British landowners of the early 19th century employed the skills of Humphrey Repton, who prepared for each client a 'Red Book' illustrating, in beautiful watercolours, the effect he intended to produce. He adapted many of Brown's creations, adding sinuous drives that provided intermittent glimpses of the house on approach and linking house and grounds with terraces and beds of ornamental flowers.

❧ REPTON'S GARDENS ☙

The many gardens designed to great effect by Repton included:

Woburn Abbey, Bedfordshire – *Notable for its flower corridor, rosary, American garden (to display new introductions) and Chinese garden, as well as its menagerie.*

Endsleigh, near Tavistock, Devon – *The land around the Tamar Valley landscaped around a cottage orné, a sporting lodge built by Jeffry Wyattville.*

Plas Newydd, North Wales – *Repton re-designed the drive and planted more trees between the house and the stables, so that on approaching the house it was impossible to see both buildings at once. The home farm was moved farther from the house to the south-west, next to the orchard.*

Ashridge, Surrey – *The garden for which the 'mixed style' was first advocated by Repton in his 1813 Red Book. After Repton's death, his proposals for 15 different types of garden were adapted and implemented by Sir Jeffry Wyattville. Features include an Italian garden, a circular rose garden, a monk's garden and holy well, an armorial garden, a conservatory, a grotto and an avenue of Wellingtonias leading to an arboretum.*

THE IMPRESSION OF ANY HOUSE, WHEN SEEN FROM A DISTANCE, IS ENHANCED BY A HA-HA

The ha-ha, a sunken wall concealed in a ditch, was a key element of the naturalistic movement of the 18th century, and became a standard feature of the English country house garden.

While it formed an effective barrier, keeping animals such as sheep and cows confined in their pastures, the ha-ha allowed uninterrupted views from the house to the park or from the park to the surrounding countryside. At houses such as Charlecote in Warwickshire the ha-ha was installed to create the illusion that deer could wander right up to the house.

The invention of the ha-ha was credited to the garden designer Charles Bridgeman, though it is likely that he had seen earlier examples in France.

Ha-has are said to be named from the exclamations of surprise uttered by those who encountered them. They vary in depth from about 2ft (60cm) at Horton House in Northamptonshire to 9ft (2.7m) at Petworth House in West Sussex.

Bridgeman was instrumental in the move away from geometric Tudor layouts, and while working on Lord Cobham's estate at Stowe he installed a ha-ha there in around 1714. Bridgeman's contribution was overshadowed, however, by that of his contemporary William Kent, painter and architect and founder of the English landscape tradition. Kent's masterpiece was Rousham in Oxfordshire, where he created vistas out towards the River Cherwell as well as woodland walks and open glades, each ending in a 'surprise' architectural feature.

By the Victorian era, reaction had set in against the landscape movement. The designers Humphrey Repton and William Gilpin argued that: 'If it be contrary to good sense to admit the cattle on the dressed lawn, it is, I conceive, equally contrary to let it appear they are admitted.' As a result, garden walls, fences and 'see-through' iron railings were erected, and the flower garden returned to the immediate environs of the house. Terraces that had been removed a century earlier were reinstated, marking the separation between house and park.

EVERY SIZEABLE GARDEN IS IN NEED OF HERBACEOUS BORDERS

The herbaceous border came into its own in the Victorian era, and the country house garden was large enough to be a perfect setting for magnificent displays.

'In many situations near houses, and especially old houses,' says William Robinson in *The English Flower Garden* of 1883, 'there are delightful opportunities for a very beautiful kind of flower border ... Here we can have the best soil and keep it for our favourites.' Among the many specimens he recommended for the mixed border were delphiniums, lilies, peonies and irises.

Wherever borders were placed in the garden – they might still form

So-called carpet bedding, using annuals en masse, was much favoured in country house gardens. At Cliveden in Berkshire in 1868, John Fleming planted such a bed depicting the monogram of Harriet, Duchess of Sutherland, using arabis, echeveria and sempervivum on a background of sedums of different colours.

elements of a parterre as well as fringing shrubberies or making borders beside grass walks – mixing colours was all-important. In the mid-19th century the amateur gardener Shirley Hibberd recommended a repetition of plant or colour groups placed at intervals along a mixed border. 'The hardy herbaceous border,' he said, in antithesis to single-species plantings, 'is the best feature of the flower garden … When well made, well stocked and well managed, it presents us with flowers in abundance during ten months of twelve … While the bedding system is an embellishment, the herbaceous border is a fundamental feature.'

By the early 20th century the herbaceous border had been not only accepted but developed, particularly as a result of the influence of Gertrude Jekyll, who also incorporated shrubby plants such as sage into the herbaceous border to provide 'filler' foliage. Her many designs, which always paid great attention to colour, included seasonal borders placed in different locations in the garden.

DECORATIVE FEATURES DESERVE A PROMINENT PLACE IN THE COUNTRY HOUSE GARDEN

The country house garden could contain statues and obelisks, garden follies and even temples.

Statues found their place in the English garden in the late 16th century and in country house gardens they were particularly enjoyed as, said John Woolridge in 1677, a 'Winter diversion … to recompense the loss of past pleasures, and to buoy up hope of another Spring'. Since marble quickly became prey to the weather, statues began to be made in lead, often painted to resemble marble or bronze.

The craze for follies really took off in the 1690s, with such eccentric constructions as the gigantic 'pineapple' built at Dunmore Park near Falkirk in Scotland. Towers were also popular, and could be climbed to admire the view. As the English landscape garden evolved, so eye-catching features were introduced to add atmosphere and interest to the 'living picture' that was the garden. In some gardens, as at Stowe in Buckinghamshire, the features were placed to mark stages in a garden perambulation. Its massive Doric arch, for instance, framed a view of the garden's Palladian bridge. At Stourhead in Wiltshire, the Pantheon is the focal point of the view from the hillside near the house.

At Dallington in Sussex, the local squire wagered that he could see the village church spire from his window. When this proved impossible, he built a conical folly known as the Sugar Loaf on his land, resembling the spire, to win his bet.

A HEAD GARDENER IS AN INDIVIDUAL OF NO LITTLE IMPORTANCE

And, says The Servants' Practical Guide, *'usually a man possessing a considerable amount of practical knowledge and a fair education'.*

The head gardener of a country house certainly had to be versatile. As well as attending to beds and borders he needed to be able to manage greenhouses, hot houses and conservatories and grow fruit and vegetables. He needed to know about landscaping and design, about soils and composts and be expert in propagating all kinds of plants, including shrubs and trees. There were also ponds, lakes and water gardens to maintain, as well as lawns, which required constant attention, especially if used for tennis and croquet (see Chapter 5).

With training as an under gardener, and with experience, the head gardener would learn how to perform each garden operation in the right season and with an eye to the weather. In spring and autumn, walls and buildings in the garden

∞ TIPS FOR THE HEAD GARDENER ∞

19th-century advice for head gardeners that still holds good today:

- *Complete every part of an operation as you proceed.*

- *Finish one job before beginning another.*

- *In leaving off working at any job, leave the work and tools in an orderly manner.*

- *In passing to and from the work … keep a vigilant lookout for weeds, decayed leaves, or any other deformity and remove them.*

- *Let no crop of fruit, or herbaceous vegetables, go to waste on the spot.*

- *Keep every part of what is under your care, perfect in its kind.*

needed to be repaired and painted, if necessary, and tools properly cleaned and sharpened. The 'basic' set of tools in the mid-19th century included: spade, shovel, rake, hoe, fork (three pronged), trowel, shears, scythe, pruning knife, hay rake, dibber, line and reel, besom and garden roller.

The intensive labour of tending the garden – digging, trenching, mowing, gravelling, hoeing and the like – was performed by a team of under gardeners. On a large estate they might number as many as 25. In common with the head gardener, the most senior of them would live in cottages in the grounds. As a perk many were given free firewood to heat their homes.

The head gardener needed a pleasant and courteous manner since he would have to show visitors around the garden. 'Take every opportunity,' said Anthony Heasel in his 18th-century guide, 'of entertaining those who come to visit your master, with a particular description of every thing in the garden, and have always some places ready for them to rest themselves on, while passing from one part to another.'

Box and yew are the most useful plants for hedging and topiary

Long used in knot gardens, fragrant box was an ideal plant for clipping into small shapes, while yew was preferred where a taller effect was required.

In Tudor times, almost every country house garden included a knot garden, created by interweaving box hedges with herbs in a beautiful pattern. Low box hedges were also used to edge beds and borders, including rose beds. To keep it looking good box needed an annual clipping in late May or early June, once it had started into growth, when topiary shapes could also be created. Care was needed to keep the foliage of other plants well clear of it, for aesthetic reasons and to maintain it in good health. Box 'balls' for formal garden designs were made by training plants or by grafting between species.

Yews were used for hedging and as specimen trees planted, for instance, in pairs at the top of garden steps. Clipping yew into shapes became the rage in the 19th century, when bird forms were especially favoured, echoing the live creatures of the dovecote.

While yew topiary was the passion of many Victorian gardeners, William Robinson thought it an abomination, declaring even unfussy clipping as a 'misuse', not only in British gardens but even at Versailles.

WATER FEATURES PROVIDE ENDLESS OPPORTUNITIES FOR HORTICULTURAL EMBELLISHMENT

As long as they were well maintained and prevented from becoming stagnant, water features were a delightful addition to the garden.

Water gardens could be anything from informal streams, bordered with flowers and trailing shrubs, to large lakes. Among the most beautiful was the canal garden at Bridge House in Surrey, designed by the landscape architect Harold Peto. Here a lily pond ornamented with hydrangeas sat alongside a pillared pergola festooned with climbers.

Any water feature needed to be placed so that it was most pleasing to the eye. 'All gardening in which water plays an important part,' wrote Gertrude Jekyll in the early 20th century, 'implies a change of level in the ground to be dealt with.' From in front of the house she envisaged that 'a flight of easy steps would descend to the Water Lily court, landing on a wide flagged path that passes all around the tank.' She recommended steps leading down, continuing below the water line, and a surrounding area planted with canna lilies.

Where there was water there could also be spouts, cascades and fountains. In hilly locations such as Chatsworth in Derbyshire, water 'staircases' were created in the early 18th century. Here, water gushed from a cascade house out of the mouths of beasts, pipes and urns, and, said Daniel Defoe, 'a whole river descends the slope of a hill a quarter of a mile in length, over steps, with a terrible noise, and broken appearance.' The design most favoured for fountains was the three-tiered *compotière*, whose shape was mirrored indoors in everything from candelabras to flower arrangements.

GARDEN FLOWERS ARE REQUIRED FOR INDOOR ARRANGEMENTS

Floral decorations were essential for the dining table, hall and other rooms in the house. Supplying them year round was the responsibility of the head gardener and his assistants.

Roses were essential for indoor arrangements and with careful selection of outdoor varieties they could be in bloom from late spring right through to autumn. Some were also forced for early spring blooming. Tulips and other spring bulbs such as hyacinths were used indoors, but even more important were lilies which, grown under glass or in a 'reserve garden', would have been available all year.

Among the other 'florist's flowers' recommended were chrysanthemums, propagated by cuttings and raised both outdoors and in, and carnations, which needed to be increased by a technique known as layering. To create large, perfect carnation blooms, circular pieces of card, each with a hole in the centre, were placed over individual buds before they opened. Hollyhocks were also grown for indoor decoration. These were raised in pots before being planted out.

> *Wide 'cutting borders' of herbaceous perennials and roses,*
> *often flanking generous paths, were created in gardens to*
> *provide flowers for the house. Here ladies of all ages would*
> *gather flowers for nosegays and pot-pourris, collect specimens*
> *for pressing and make sketches. While the head gardener*
> *welcomed the mistress of the house or her daughters into the*
> *garden to choose flowers for arranging, or fruit such as grapes*
> *for embellishment, he would always prefer to pick them*
> *himself. Ladies who valued their gardeners were advised to*
> *'humour this weakness'.*

THE SUNDIAL IS AN APPROPRIATE ORNAMENT FOR THE GARDEN

As well as having a practical use, sundials became popular garden ornaments in country house gardens from the late 15th century. Many were furnished with mottoes, usually in Latin, reflecting on the passing of time, and with family crests and other ornaments.

The move of sundials from church walls into large gardens, where they became constant reminders of the cycle of days, months and seasons, came about as a result of Renaissance advances in accuracy and design. Henry VIII set the trend with more than 20 horizontal and beautifully ornamented sundials ordered by his royal horologer, the Bavarian astronomer and mathematician Nicholas Kratzer.

The typical sundial of the period consisted of a metal plate on which was

Even in the Victorian age, when every country house had many accurate clocks, sundials continued to ornament the garden. In an era that combined practicality with religious fervour and sentimentality, the use of traditional sundial inscriptions remained popular.

mounted the gnomon – also made in metal – the arm that cast the shadow to mark the hour. The whole was then mounted on a stone pedestal. At Wollaton Hall in Nottinghamshire Robert Smythson installed a massive sundial as the central feature in a parterre. Although the original garden no longer exists there is evidence that the garden also had a separate area devoted to a large sundial collection.

⌖ SUNDIAL SAYINGS ⌖

Latin was traditionally used for sundial inscriptions. Many garden sundials took their mottoes from those originally used on churches.

Lex Dei lux Dei	*The law of God, the light of God*
Fugit irreparabile tempus	*Irredeemable time flies away*
Umbra sumus	*We are shadow*
Noli confidere nocti	*Trust not the night*
Vulnerant omnes, ultima necat	*Every (hour) wounds, the last kills*

A PERGOLA WALK SHOULD BE IN KEEPING WITH THE MATERIAL OF THE HOUSE

A key concept of 19th-century design for the country house garden and one espoused by such luminaries as Gertrude Jekyll and Edwin Lutyens.

Tunnel-arbours had been popular features of country house gardens in the 16th century but were revived with enthusiasm towards the end of the 19th, using as supports everything from sculptured Doric and Ionic columns derived from Italian gardens to timber (especially oak) and pillars of stone, brick or chalk. For more delicate effects, metal frames, originally imported from Germany, were also employed. Simplicity was key. 'On no account,' said William Robinson, should one 'let the "rustic" carpenter' adorn it with 'the fantastic branchings he is so fond of'.

A pergola swathed in scented plants was a place for home owners and their visitors to stroll at leisure. In *The English Flower Garden* of 1883 Robinson described the climbing rose 'Gloire de Dijon' as 'the most precious flower that ever adorned the garden', and also recommended vines as 'living drapery' for a pergola, together with wisteria, Virginia creeper, honeysuckle, jasmine and clematis.

At Aberglasney in Wales it is still possible to walk through the massive, darkly romantic yew tunnel planted in the 18th century, while at Moseley Old Hall in Staffordshire an arbour of hornbeam leads to a tunnel of oak covered with *Clematis viticella* and *C. flammula* and with the claret vine *Vitis vinifera*. The design is based on one illustrated in Thomas Hill's *The Gardener's Labyrinth* of 1577.

THE CONTINUOUS SUPPLY OF FRESH VEGETABLES IS A VITAL PART OF THE GARDENER'S DUTIES

Behind the high walls of the country house kitchen garden, all manner of vegetables were grown in succession to keep the house well fed throughout the year.

Until the early 18th century, flowers and vegetables tended to be planted together in parterres and 'pleasure gardens' to create decorative effects, but from then on they became separated, and vegetables were often grown on a large scale. The planting plan was devised to allow for crop rotation but also to ensure that the crops needing the most warmth, such as winter lettuces and early carrots, were provided with the sunniest, most sheltered positions. Members of the cabbage family were given particular attention in the kitchen garden, as were peas, beans and spinach.

Herbs had an important place in the kitchen garden, usually being placed in a border that was easily accessible from one of the paths within the garden. These would include parsley, thyme, tarragon, fennel, marjoram, rosemary, sage and bay. Between the herbs, shallots and small onions were planted.

For forcing outside the greenhouse, glass cloches were placed over vegetables such as salads.

Peas and beans are the most important of all kitchen garden crops

Both these crops were greatly prized at the country house table, but special efforts were made to produce early crops of peas.

The garden pea reached Europe from the East in the 16th century and immediately became dubbed 'the Prince of the vegetable garden'. A taste quickly developed among the well-to-do for eating them green and immature, rather than as a floury dried vegetable. Country house gardeners across England strove to have their peas ready for the table by 4 June, the birthday of King George III, and a crop was even grown for the Christmas table by planting them in pots in the autumn then raising them in a heated greenhouse for the next couple of months.

The beans most commonly grown in country house gardens were varieties of broad beans, which were podded and eaten as a valuable source of protein, although for fine dining very small beans were required. As with peas, successional sowing was necessary to extend the season. Brown, red and green Windsor were among the most popular varieties. Also popular – and versatile – were kidney beans, which could also be eaten young, and green French beans. Remarking on their

The scarlet runner bean, introduced to Britain by John Tradescant the Elder, gardener to Charles I, was originally grown for its flowers. For the earliest crops it was the practice to dig up the roots following the harvest, store them in a frost-free place then replant them in March.

culinary value, Robert Thompson says that: 'If the green pods are superabundant in summer, they may be preserved in salt, for use in winter; they may be made into a pickle alone, or together with other vegetables; and, finally, the ripe seed can be used in a variety of ways, in haricots, soups, and stews.'

WALLS PROVIDE A PERFECT PROTECTION FOR FRUIT TREES

The walled kitchen garden, which gained great popularity in the 18th and 19th centuries, was the perfect way to protect crops and was particularly suited to the growing of espaliered or cordoned fruit trees.

As Sir Robert Stapleton, a 16th-century Yorkshire landowner, wrote:

For growing tender Mediterranean fruit such as peaches and the apricot 'Moorpark', a highly fragrant variety, and for cultivating outdoor vines, south-facing walls were not only made extra thick but were heated by fireplaces built into their backs. Heat was conducted through flues set into the brickwork.

'If you defend not your orchard from the northerly and easterly wind you can never have dainty fruit … It is convenient to defend your orchard from the wind on all sides, otherwise your apples & other fruits will be cast down often times ere they be ripe.'

In a large walled garden, the traditional arrangement was to divide it into four, with each quarter again divided into four and with a fruit tree in the centre. Apples and gooseberries might be planted around the borders of each quarter, with espaliered or cordoned fruit against the high walls, including apples, pears, plums, quinces, nectarines and cherries. The walls not only served to keep out rabbits, deer and other 'pests' but, because they warmed up during the day and released their heat gradually during the night, they kept fruit at an even temperature, which encouraged both development and ripening.

RARE AND UNUSUAL PLANTS MAY BE KEPT IN A CONSERVATORY

The conservatory, which could either be freestanding or attached to the house, offered the chance for growing exotic specimens of all kinds.

From the early 18th century, when conservatories were first built for the wealthiest country houses, construction was from brick or stone and designs were rectangular. Glass panes – at that time extremely costly – were small and numerous.

Conservatories became hugely popular following the completion of Joseph Paxton's Great Conservatory at Chatsworth in 1841, built specifically to house the giant Amazonian water lily, and his Crystal Palace, constructed for the Great Exhibition in 1851. The Victorians favoured hexagonal or round shapes with steep roofs; panes were large and design details included glass fanlights, bay windows, arched doorways and elegant cast iron eave brackets.

The conservatory often opened straight into the drawing room or might serve as a means of disguising the servants' wing of the house, with an opening into the hall. Alternatively, as at Mentmore in Buckinghamshire, it might link the morning and smoking rooms.

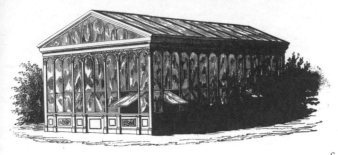

Inside the conservatory, favourite plants included abutilons, jasmines, fuchsias, heliotropes, palms and ferns of all kinds including maidenhair, holly and Boston ferns. Within this 'jungle', which became ever more exotic as plant hunters brought new species from distant lands, might be a pool with water ferns, water lilies and fish, and even a fountain. Chairs, tables and benches were of cast iron, designs favoured by the Victorians mimicking the shapes of tree trunks and roots with 'foliage' embellishments.

Additional heat for the conservatory was provided by a stove. The gardener needed to know whether the plants demanded dry or moist heat, and adjust the system accordingly. 'Bottom heat for stoves,' advised Robert Thompson in *The Gardener's Assistant* of 1859, 'is sometimes supplied by means of tanners' bark placed in a pit in the body of the house, and on this the pots containing plants are set, or occasionally plunged in it, to a greater or lesser depth, as when it is necessary to excite their roots more than their tops. But generally,' he concludes, 'a fair share of bottom heat is maintained by placing the pots on slate or stone, under which there is a heated chamber, whether by flues, hot-water pipes, tanks, or heated air; but hot-water pipes are the most eligible.'

HEAT IS ALWAYS REQUIRED FOR GROWING EXOTIC FRUIT

A saying particularly true for oranges but also for prized exotic fruits such as pineapples, which were cultivated at country houses.

Early orangeries, such as the one at Dyrham Park in Gloucestershire, built in 1702, originally had solid roofs, but these were replaced by glass a century later. Courting couples who wanted to be together but out of sight relished the confines of the orangery for clandestine meetings.

As well as oranges, lemons and limes were grown in the orangery, while vines, peaches, apricots, plums, figs and cherries were cultivated under glass in large houses with sloping glass roofs reaching down to ground level. Peach trees might be mixed with vines. Both figs and cherries were planted in pots, although figs might also be set directly into a border.

For early pickings of redcurrants, gooseberries and raspberries, plants were potted up in late winter and brought on in heated glasshouses. For strawberries at Christmas, plants were propagated then raised in pits in the same way as pineapples.

For protecting pineapples and melons – and out-of-season vegetable fruits such as cucumbers – elaborate pits were constructed, heated by hot water pipes from below and by the sun from above, entering through a sloping glass roof. Beneath the plants was a chamber into which rotting manure from the stables was inserted to 'supply moisture and ammonia', which also helped to deter insect pests. Plant roots might also be allowed to penetrate – and benefit from – this rich compost. Good ventilation was essential, particularly for fruits such as plums and cherries needing insect pollination.

To ensure a supply of fruit early in the season, trees needed heat to coax them into life in mid-winter. Once the buds broke constant attention was necessary, for, as the 1862 *Book of Garden Management* said: 'From the time the bud bursts its horny sheath until the luscious fruit melts in your mouth, all work and no play – all growth and no check, must be the stern regimen of the successful cultivator.'

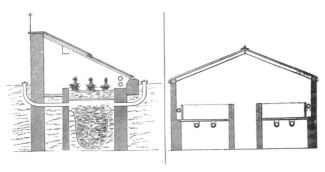

AN ICE HOUSE MUST BE SITUATED IN A SPOT SUFFICIENTLY ELEVATED TO ALLOW COMPLETE DRAINAGE

A feature of country house estates from the 17th century, ice houses needed to be constructed so as to keep the contents usable for months on end.

The side of a north-facing slope was the ideal position for an ice house, with a door facing south-east. The alternative – as in the earliest examples – was to dig a pit and line it with bricks. In either case, drainage needed to be provided for water to run away through a grating, plus an air trap to prevent warm air getting in and melting the ice. Damp was even more of a problem, especially in ice houses constructed on water-retaining clay and loam. Those built on free-draining chalk and gravel were undoubtedly the most effective. Keeping the house away from trees was also important because although they provided shade their roots were a source of damp.

Good insulation was vital. This was provided by constructing double walls, with insulation such as sawdust between them, and by covering the roof with straw thatch. On top of this might be placed earth planted over with ivy. Barley straw, put into canvas bags to make it easier to manage, was also used in layers between two or three doors.

Filling the ice house was an annual task, but it was topped up as the opportunity arose. Until the 1800s, ice was collected from lakes on the estate and from specially constructed shallow 'freezing pools'. In milder areas, compacted snow was commonly used in its place. Robert Copeman, the steward of Blickling Hall in Norfolk, recorded in April 1790: 'There was a great fall of snow this afternoon, as soon as ever it was thick enough I spoke to the Gardeners and had as many Labourers as I could. The ice house is about half full; the men kept at it between 9 & 10 o'clock.'

> *Food was sometimes stored in the ice house, but more usually the ice was removed as needed and put into ice boxes and chests in the kitchen, where it was used for cooling ices and other cold dishes.*

From 1842, when the first shipment of ice from Fresh Pond, Boston, arrived in Britain on the barque *Shannon*, a better, purer form of ice became available. The Wenham Lake Company set up offices in London and organized the distribution of this superior ice to estates around the country. Quoting prices, *Cassell's Household Guide* says that: 'American ice is sold by many ice dealers at about 7s. per 100lbs.; 3s 6d. per 50lbs.; 2s per 25lbs.; and in any smaller quantity at 2d per lb.; and is packed in a mat or blanket, for the country, at a further charge of 2s per 100lbs; and in larger quantity at some reduction in this rate.' Later, Norwegian ice also became available. Though considered poorer in quality it was available through the summer months when most needed.

BREWING BEER IS EXCLUSIVELY CONFINED TO HOUSEHOLDS IN THE COUNTRY

Many country establishments had brewhouses specifically for this purpose, while in smaller ones beer was made in the scullery, with laundry tubs doubling as brewing tuns.

The tradition of brewing beer grew up in monasteries and convents, to supply refreshment to pilgrims and travellers, and the oldest country house brewery, at Lacock Abbey in Wiltshire, resulted from the purchase of a nunnery by Willian Sharington from Henry VIII. Water heated by the furnace in the adjoining bakehouse (from where yeast was supplied) was boiled and then cooled before being run off into the mash tun containing barley and malt.

Next, the mixture was stirred for several hours with a mash paddle to help convert the starch in the cereal into sugar. The liquid subsequently drawn off was boiled in the copper with sugar and hops before being drained yet again, this time into a fermenting tun where yeast was added. In just a few days a good brew was produced, which, after skimming, was either drunk at once or put into casks to mature further. Generally three strengths were brewed: a weakish thirst-quencher for everyday consumption, known as beer; a medium strength to accompany family meals, known as ale and stored in casks, possibly for a year or more; and a strong brew – malt liquor, which was invariably bottled – for special occasions.

Malting the barley for beer making was done in the winter, possibly in a separate malt house situated alongside the brewhouse, or by a local miller. Malting involved steeping the grain in water then leaving it to 'sweat' before spreading it out to dry. Soft water such as that found in Devon, Nottinghamshire and Yorkshire, produced the best beers. To keep the yeast alive it was kept warm and stored in tubs provided with regular supplies of oxygen. Hops for flavouring were grown in a hop garden on the estate or brought in from elsewhere. Chamomile flowers were also used for flavouring beer, adding pleasantly to its bitter taste.

For storing beer, either casks or bottles were used, and these were kept in a beer cellar maintained at a constant temperature of 60°F (15.5°C). To aid transport from the brewhouse, beer might be piped into the cellar, as at Chatsworth in Derbyshire and Uppark in West Sussex. While the butler retained overall responsibility for brewing, a large house would employ a brewer. Other staff involved in brewing included housemaids, dairymaids, coachmen and gardeners. Casual, local labour might also be brought in at busy times, particularly March and October – the best brewing months.

No dovecote can possibly survive if rats have found an entrance to it

One very good reason why dovecotes have long been placed well above ground. The oldest designs are round or shaped like beehives, giving easy access to both birds and eggs.

Building a dovecote, which might house hundreds or even thousands of birds, was originally a privilege of the lord of the manor and forbidden to his tenants – leading to many complaints, since the birds were free to ravage the crops of the tenant farmers. The doves and pigeons kept in the dovecotes were an important source of meat and eggs, particularly in winter when other fresh ingredients were hard to come by. The young birds known as squabs or squeakers were the greatest delicacies.

The distinctive cupola set atop many dovecotes, often equipped with landing ledges and sheltered perches, made it easy for birds to go in and out. Within, internal shutters might be fitted, or even dormer windows. Simpler versions had ledges placed around the exterior in front of the entrance holes. Small compartments within the cotes ensured ideal nesting conditions. Each

⟳ Ancient praise ⟲

Describing the manor of Owen Glendower at Sycharth in North Wales the 14th-century poet Iolo Goch wrote the lines:

Fine mill on smooth-flowing stream;
Dovecote, a bright stone tower;
A fish-pond enclosed and deep,
Where nets are cast when need be.

In country house grounds, dovecotes were often placed near poultry and close to the bakehouse, brewhouse and stables. Since the birds needed plentiful supplies of water the cotes were also put close to fish ponds. To encourage birds to stay in a new dovecote it was the practice to sprinkle the floor with a strong-smelling substance such as asafoetida, a pungent spice related to fennel.

pair of birds might, in a lifetime of seven years, produce two or more chicks twice each year.

The manure from the dovecote, prized as a fertilizer, was removed twice a year, in November and February. Hygiene was vital since, as *Cassell's Household Guide* advises: 'The young of the dove-cote pigeon, like all others of the columbine order, are reared in a nest lined by their own dung, which if left in the hole after the birds are gone is apt to harbour vermin ...' which will 'destroy every young pigeon within their reach.'

THE LAND STEWARD SHOULD HAVE NO OTHER OCCUPATION

Not least because his was a role of great responsibility within the estate. He was the top man, on whom his master relied absolutely.

Any man wishing to fulfil the role of land steward needed to be both versatile and totally reliable. As well as knowledge of agriculture he needed skill in accounting, surveying and architecture, plus the personal accomplishments essential for dealing with 'inferior servants' and with the estate's tenants.

The first job of the newly appointed land steward was to survey the estate and make a detailed inventory of everything in it. 'From this survey,' said *The Complete Servant*, 'regular memorandums should be made in a book, of every thing necessary to be remarked or executed, of the places where deficiencies are found, or improvements may be made; of buildings and repairs necessary;

insurances, dates of leases, rates, nuisances, trespasses, live and dead stock, game, timber, fencing, draining, paths and roads, culture, commons, rivers, and sea coasts, and of every other specific article relative to his trust, which deserves attention, and therefore ought not to be committed to loose papers, or left to memory.'

The good land steward kept a day book or journal, a ledger, a memorandum book and a general inventory. He also supplied those who worked for him with account books and examined these at regular intervals.

As each farm on the estate was surveyed, the land steward would provide tenants with a copy of their map. The steward also needed to keep a record or 'terrier' of all common lands. Any boundary disputes were settled by a jury at the manor court, and also recorded. In order to keep a watchful eye on such matters, and on repairs and renovations (including those necessary to roads and bridges and the like), the conscientious steward rode around the estate daily.

∝ THE GOOD STEWARD ∝

Additionally the steward of excellence would:

- *Encourage improvements in cultivation and husbandry, including encouraging tenants to plant orchards.*

- *Manage woodlands and the planting and felling of trees.*

- *Look out for the presence of valuable minerals on the estate.*

- *Let as much of the estate land as possible, and provide leases for tenants.*

- *Improve the value of the estate by adding roads or facilities such as fisheries.*

- *Collect rents and make sure that the money was taken to the bank for safe keeping.*

CARE OF DEER IS ENTRUSTED TO BOTH PARK-KEEPER AND GAMEKEEPER

A vital job on estates where deer were kept. Rearing of pheasants for shooting, and management of shoots, were also within the remit of these roles.

Summarizing the duties of the park-keeper, Giles Jacob in *The Compleat Sportsman* of 1718 says that he must '… daily take turn around his Park, and keep a constant Account of the Number of his Deer; and oftentimes watch them at Night, for their Preservation against unlawful Hunters, especially in Moon-shiny Nights and the Rutting Season. He must take care to calculate an exact Number of Bucks and Does proper to be kill'd in each Season … and at the same Time not to over-stock the same, preserving a proper number of young Fawns to be bred up in the steads of those he kills …'

Sound knowledge of the law was essential to the position of gamekeeper. In 1671 an act was passed that not only granted the right of every landowner to appoint a gamekeeper but allowed him to confiscate dogs, firearms or other 'implements of the chase' from anybody suspected of being unqualified to hunt and to bring any poachers he apprehended to face local justice. The man appointed to the post might also act as the park-keeper or land steward, or he might be a local tenant farmer or neighbouring estate owner.

Women could be gamekeepers too. One of the best known is Polly Fishbourne, who served on the estate at Holkham Hall in Norfolk in the early 19th century. This formidable woman, with her hair cut short and topped with a man's hat, was renowned for her strength.

Work on estates was often a family affair. In 1885, for example, *The Field* magazine ran advertisements offering both gamekeepers and their sons for hire. A keepers' register was also kept at a London gunmaker's. Cottages on the estate were supplied for the park-keeper and if necessary for the head and second gamekeepers.

THE DAIRYMAID ATTENDS TO THE POULTRY

Her duties also included raising chickens, plucking them ready for trussing and making sure that, in summer, eggs were preserved for winter use.

The poultry cared for by the dairymaid included ducks, geese, turkeys and possibly guinea fowl as well as chickens. All were kept in a poultry yard close to the brewhouse or another building to provide some warmth, which would have helped the birds to thrive.

Hen houses would also be provided, as colourfully described by Gervase Markham in his *Perfect Husbandry* of 1615 (here in modern spelling): 'Your

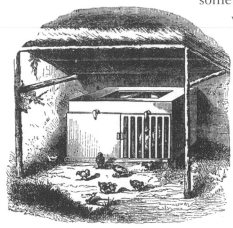

hen house would be large and spacious with somewhat a high roof and walls strong and windows upon the same rising; round about the inside of the walls upon the ground would be built large pens of 3ft high for Geese, Ducks and Fowls to sit in.

'Near to the eavings …. would be perches … on which to sit your cocks, hennes, capons and turkeys, each on several perches as they are disposed … Let there be pins stucken into the walls so that your poultry may climb to their perches with ease, let the floor be ... of earth smooth and easy, let the small fowl have a hole at one end of the house to come in and out at when they please ...'

The dairy maid gathered up the eggs once a day, and needed to be in tune with the ways and needs of her birds, for as Markham pointed out, 'Some hens will by cackling tell you when they have layed, but some will lay mute, therefore you must let your own eye be your instruction.' To help fatten geese and turkeys for good eating, and to keep birds of all kinds laying through the winter, they might be fed malt or mixtures such as oatmeal and treacle or barley meal mashed with milk. If wheat grains were given they first needed to be crushed and soaked in water.

BUYING AND SELLING CATTLE IS THE BUSINESS OF A BAILIFF

Just one of his many duties on the home farm. The bailiff would also, under the instruction of the land steward, administer parts of the estate let out to tenants.

The medieval bailiff was a freeholder who owned his own land; it was his role to allot jobs to the peasants while taking care of running repairs to buildings, for which he hired skilled labourers such as carpenters and blacksmiths. In time the bailiff's role extended to that of farm manager, and the bailiff of a country house estate needed extensive knowledge of agriculture. It was his task, said Giles Jacob in his *Country Gentleman's Vademecum* of 1717, 'to buy and sell Cattle and Horses for the Plow, direct and order Plowmen, buy Corn for Seed, inspect Plowing, Sowing and all manner of Husbandry Affairs'. He was also expected to keep accounts of all cattle bought and sold.

Up to the 18th century, the bailiff might also be expected to take on the role in the house that was later fulfilled by a butler. At Burley in Rutland, for instance, the Earl of Nottingham demanded that his bailiff: 'Attend and wait at table and also in ye Hall' when guests were being entertained. However, the bailiff would not be required to wear livery (see Chapter 1).

As well as controlling the activities of tenant farmers, the bailiff needed to make sure that land was not overrun by moles or woods with wild pigs. He had to ensure that cattle and dogs did not stray and punish offenders as appropriate, and he needed to protect the land so that its valuable crops and manure were not stolen.

THE GROOM SHOULD BE AT THE STABLE AT SIX IN THE MORNING

And an hour earlier than this in the summer, or at any time when his master wished to ride before breakfast.

Timekeeping aside, the most important quality in any groom was his ability to attend to the horses in his care throughout the day. As *The Complete Servant* says, horses '... regularly managed, under the humane superintendence of a diligent and conscientious Groom or Coachman, will have healthy and beautiful appearance, and in a great measure escape from many diseases to which they would otherwise be liable.'

Horses were given water and feed each morning, then cleaned, first with a curry comb, to loosen any dust and dirt, then with a whalebone brush. Finally they were rubbed down with a woollen rubber or a clean cloth. Next came the more intimate care: 'The horse is then turned round in the stall, and his head is next brushed well and wisped clean and smooth, with a damp lock

⟪ HORSE CARE ⟫

Tips for grooms from The Complete Servant:

'The finest-conditioned horses in England are fed thus: when at grass, equal quantities of oats and old white peas; when in the stable, two thirds oats and one third old white peas.'

To test oats for quality
'The shorter and fuller the grain the better; when bitten in two they should be dry and mealy: they should feel hard in the hand, and when hard grasped they should slip through the fingers; oats with long bodies and tails are the worst.'

To treat sores and bruises
'Over the whole sore, or where a part is bruised, or where there is a tendency to suppuration, a poultice should be applied and kept on by suitable bandages. The poultice may be made of any kind of meal, fine bran, bruised linseed, or mashed turnips, carrots, &c.'

Ointment for wounds
When the wound *'begins to put on a healthy appearance, and seems to be clean and of a reddish colour, not black or bloody; then there may be applied an ointment made of tallow, linseed oil, bees' wax and hog's lard, in such proportions as to make it of a consistence somewhat firmer than butter.'*

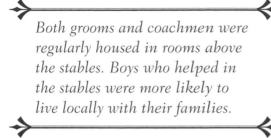

Both grooms and coachmen were regularly housed in rooms above the stables. Boys who helped in the stables were more likely to live locally with their families.

of hay.' The guide continued 'After this, his ears are drawn through the hands, for several minutes, till made warm, and then the insides of his ears are wiped out with a damp sponge ... [it] is then applied to the eyes ... The head is afterwards finished by rubbing it with a cloth ... The Groom next examines the horse's heels, picks out the dirt from the feet, and washes its heels, with a water brush and plenty of water.'

Horses' feet and legs needed special attention. They had to be washed and rubbed dry and the feet dressed if they showed any sign of trouble. If a horse had been ridden hard and its feet were too hot, they would be treated with a potent 'stopping' mixture made by mixing 'equal quantities of cow-dung, clay, tar, and kitchen grease with urine, to the consistency of a stiff paste', which was kept handy in a tub in the stables. Bandages were also applied to a weary animal's legs between the knees and the fetlock joints to prevent swelling.

A HEAD COACHMAN'S OFFICE IS ONE OF CONSIDERABLE TRUST

This was particularly true in smaller houses where there was no master of the horse or clerk of the stables to whom he had to report.

When in sole charge, the head coachman would not merely drive the family carriages but direct the grooms, postilions and stable hands, order and buy all the hay, corn and other necessities for the horses, and look after the coaches themselves, making sure that they were kept in tip-top condition. Knowledge of farriery was also essential. Most of all, any coachman, whatever his status, needed to be good with horses and dexterous with the reins. 'Where two

coachmen and as many grooms are kept,' said *The Servants' Practical Guide*, 'the carriage is probably out three times a day; where one coachman and a groom are kept it is usual to have a carriage out twice only, a pair of horses in the afternoon, and a pair or single horse in the evening, or a pair in the morning and again in the afternoon. A horse for night work,' it adds, 'is frequently kept when the carriage is much required in the evening, and when the condition of the carriage horses is considered.'

⚬ Spick and Span ⚬

Cleaning tips for coachmen from The Complete Servant:

Black Dye for Harness
'The colour of harness that has become rusty or brown by wear, may be restored to a fine black after the dirt has been sponged and brushed off, by using the following mixture: Boil logwood chips in three quarts of soft water, to which add three ounces of nut-galls [galls from hazels], finely powdered, and one ounce of alum; simmer the whole together for half an hour, and it will be fit for use.'

Liquid Blacking for Harness
'Take 2oz of mutton suet, melted, 6oz of purified bees' wax, melted; ¼lb lamp black; 1 gill of turpentine; 2oz of Prussian blue, powdered; 1oz of indigo blue, ground; 6oz of sugar-candy, melted in a little water; and 2oz of soft soap. Mix, and simmer over the fire 15 minutes, when add a gill of turpentine. Lay it on the harness with a sponge, and then polish it.'

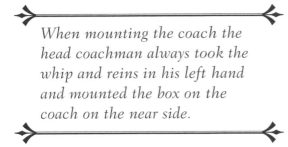

When mounting the coach the head coachman always took the whip and reins in his left hand and mounted the box on the coach on the near side.

The head coachman needed to rise early to supervise the care of the horses. Following breakfast in the house, all staff would return to the stable, 'shake down the litter on each side of the horses and put the stable in good order, in expectation of their master, who probably pays them a visit after breakfast to inspect the horses, give orders, or make enquiries.' Then followed the cleaning of harnesses and other tack, supervised by the head coachman, and the cleaning of coaches. Each day they needed to be washed and the springs, straps and any other leather trappings blacked. The wheels were thoroughly greased and oiled, the inside of the coach brushed, the glass polished and the lamps cleaned and trimmed.

THE CHAUFFEUR MUST KEEP HIS CAR IN PROPER CONDITION

With the arrival of the motor car, the uniformed chauffeur became an essential member of the 'outdoor' country house staff.

As motoring gradually took over from the use of carriages and traps in the early part of the 20th century, so chauffeurs replaced coachmen and footmen. Large houses with substantial means might employ separate chauffeurs for the master and mistress of the house and for other family members. Chauffeurs needed to be expert in all the workings of the motor car and its repair. Stable blocks were converted into garages to house vehicles such as the early Daimlers, Rolls Royces and Mercedes, as well as cars like the Clément-Talbot, first made from imported French parts and developed by the 20th Earl of Shrewsbury and Talbot, a pioneer of motoring, whose principal country residence was at Ingestre Hall in Staffordshire.

Motoring made travelling to and from London easier and more flexible but could play havoc with the tightly knit arrangements of a household. No longer, for instance, did hostesses know to the minute (as printed in railway timetables) exactly when guests would be arriving. And unchaperoned drives were a perfect opportunity for men and women to become intimate.

Women quickly took to motoring – as drivers as well as passengers. *The Lady's Realm* of 1904, reporting on The Ladies' Automobile Club, recorded for example that the Countess of Kinnoull was 'an all weather motorist' and that: 'In her 14 h.p. Chenard & Walker car which ranked among her wedding presents she recently motored from her charming Highland home – Dupplin Castle, Perth – to London.' On a sartorial note it added: 'Lady Kinnoull dresses simply and daintily on the car, preferring a three-quarter jackal fur coat for fine and a leather coat for wet weather, a small toque and motor veil, short skirt and thick shoes, which she regards as easier to use than boots for the foot work involved in driving.'

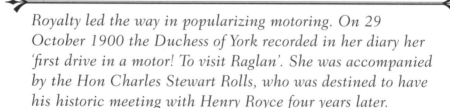

Royalty led the way in popularizing motoring. On 29 October 1900 the Duchess of York recorded in her diary her 'first drive in a motor! To visit Raglan'. She was accompanied by the Hon Charles Stewart Rolls, who was destined to have his historic meeting with Henry Royce four years later.

INDEX

Aberglasny 183
Adam, Robert 19, 147
address, terms of 108–9
afternoon tea 66–7
Albert, Prince 25, 164
Alexandra, Queen 121
Amundeville, Lord/Lady 129
Antony House 146
Ashridge 171
Austen-Leigh, Edward 159

bacon 101
bailiffs 198–9
Baker, John 166
balls 117–19, 130, 159–60
    servants' 139, 162–3
    suppers 160–2
beans 185–6
Bearwood 25
Beauclerk, Lord Frederick 145
'bed-hopping' 126
bedrooms 30–1, 36–8, 47–8
beer brewing 191–2
Beeton, Mrs 7, 11, 15, 27, 42–3,
    45, 55, 57, 66, 69, 71, 73, 80,
    94, 98–9, 104, 109, 111, 118,
    130, 158, 163
bells 29
beverages 102
billiards 21–2, 117–8
blackleading 42, 43
Blenheim 24–5
Blickling Hall 190
bonnets 73
Bowood 32
bread troughs 87–8
breakfasts 49–50, 80, 104
Bridge House 179
Brown, Lancelot 'Capability' 170
Bullet Pudding 164
Burghley 35
Burley 199
butlers 17, 41, 49, 54–5, 68, 75,
    96–7, 108, 124, 131, 133–4,
    137, 140, 150–1, 160–2, 192
butter making 86–7
Byron, Lord 129

Cannons 64, 156
card games 154–5
Cardiff Castle 22, 25
Carew, Richard 146
Carlshalton House 24
Cassiobury Park 142
cattle 198–9
Chandos, Duke of 64, 156
chaperones 126–7
charades 152–3
Charborough 21
Charlecote 172
Charles I 185
Chatsworth 24, 122, 180, 187,
    192
chauffeurs 124, 203–4
children 36–7, 41, 60–3
Christmas 163–5
cigars 133–4
cleaning 39, 42–5, 47–8, 51–3,
    75, 202
Cliveden 173
clothing 62, 112–14, 121
    lady's 70–3, 109–10, 112–14,
    116–17, 118, 127–8,
    140–1, 144, 204
    men's 69–70, 113, 117,
    118–19, 141, 144
coachmen 14, 160, 199–203
conservatories 187–8
conversation 124–5, 133, 134–6
cooking techniques 89–91
cooks 17, 49–51, 79–82, 93–4,
    104, 108, 137
    'man cooks' 79, 108
Copeman, Robert 190
copper pots 84
Cragside 25
cricket 145
Croome Court 170–1
croquet 142–3
cutlery 51–3
cycling 113–14

daily routines 40–75
dairymaids 86–7, 197–8
Dallington 175

dancing 117–19
days off 165–6
deer 196–7
dinner 80–2, 105, 110
    courses 128–9
    end of the 132–3
    royal meals 122
    service 130–1
discipline 11
dovecotes 193–4
drawing rooms 18–19, 66–7,
    110–11, 132–3
Dunmore Park 175
Dyrham Park 188

Edward VII 121, 150, 154
electricity 9, 25–6
embroidery 158
Ensleigh 171
entertainment 113–16, 138–67

family matters 134–6
finances 9, 13, 15
fish kettles 90–1
Fishbourne, Polly 196
fishing 146–7
flowers 173–4, 180–1
follies 175
food temperature 95–6
footmen 14, 41, 43–5, 49, 52,
    65, 67–9, 75, 108, 110–11,
    123–4, 134, 150, 164, 166,
    203
friendships, staff 166–7
fruit trees 186–7, 188–9

gadgets 52–3
gambling 139, 145, 154–5
game larders 92–3
garden parties 140–1
gardeners, head 176–7, 181
gardens/grounds 168–204
garnishes 100
George III 185
Gilpin, William 173
gloves, long 127–8
Goch, Iolo 193

Goodwood House 167
governesses 41, 62–3, 108
gravy 94
Greville, Charles 10–11
grooms 199–201
guest rooms 30–1

ha-has 172–3
hallboys 51–2
halls 16–17, 136–7, 165
ham 101
Ham House 19, 33
Harleyford 78
Hatfield House 25–6, 35,
    143–4, 162–4
heating 9, 22–3
Heaton Hall 35
hedging 178
Hemingford Park 145
Henry VIII 181, 191
herb gardens 184
herbaceous borders 173–4, 181
Hervey, Lord 152
Hibberd, Shirley 174
Holkham Hall 12, 196
horses 199–203
hounds 117
house stewards 13, 15, 108, 137
housekeepers 15, 17, 48–51,
    102–3, 108, 137, 163
housemaids 75, 108, 110, 124
Humphry, Madge 123–4
hunting 70, 112, 116–17, 150–2

ice houses 140, 190–1
Ingatestone Hall 88
Ingestre Hall 203
ironing 59–60

Jacob, Giles 196, 198–9
Jekyll, Agnes 66–7
Jekyll, Gertrude 174, 179, 183
jelly 98–9

Kedleston Hall 21, 146–7
Kent, William 172–3
ketchup 103
Ketteringham 108
Kingston Lacy 31, 63
Kinnoull, Countess of 204
kitchen maids 49–50

kitchen odours 23, 78
kitchens 76–105
Knight, Fanny 164
Knole 145

Lacock Abbey 191–2
lady's footman 65
lady's maids 17, 41, 49, 70–3,
    75, 108, 120, 137
land stewards 194–6
landscaping 170–1
Lanhydrock 88
laundry 56–60, 73
leisure 138–67
Lennox, Lord William 145
libraries 32–3
lighting 25–6, 74–5
livery 14
Longleat 14
Losely 32
Loudon, J.C. 18–19, 91
love 126–7, 139, 165, 167,
    188–9
Lowther Castle 37–8

Manderston 29
mangles 59
manners 17, 106–37
Markham, Gervase 197–8
Marshall, Agnes B. 91, 97, 100,
    104
master of the house 10–11,
    69–70
master-servant relationship
    10–12
mealtimes 49–50, 61–2, 64,
    66–9, 80–2, 104–5, 121,
    149–50, 160–4
meat 89–90
mistress of the house 11–12,
    70–3, 110–11
Morris, William 19, 37
Moseley Old Hall 183
motoring 112, 113, 203–4
music 18, 156–7, 160, 165
musical evenings 156–7

names 108–9
Nottingham, Earl of 199
nurseries 36–7

nursery staff 60–2
nurses 27, 28

orangeries 188–9
Osbourne House 25
Owen Glendower 193

park-keepers 196–7
Paxton, Joseph 122, 187
peas 185–6
Penshurst Palace 33
Pepys, Samuel 33–4
pergolas 183
Plas Newydd 171
Pollnitz, Baron de 123
port 133–4
Portland, Duchess of 166
Portland, Duke of 26
portraits 33–4
poultry-keeping 197–8
prayers, family 45–7
preserves 102–3
'Pug's Parlour' 136–7

receiving visitors 119–22, 166–7
refrigeration 101–2
Repton, Humphrey 92, 170–1,
    173
roasts 89–90
Robinson, William 7, 173, 178,
    183
Rousham 173
royals 204
    suites 37–8
    visits 121–2
Russell, Anna, Duchess of
    Bedford 66
Rutland, Duke of 10–11

Sackville-West, Victoria 126
salads 96–7
Saltram House 19
Sandringham 150
sandwiches, garden party 141
sanitation 9, 24–5, 36
sauces 93–4
savouries 105
scullery maids 42–3, 75, 85
servants' halls 16–17, 136–7, 165
servants' quarters 31, 34–5

shoes/boots, cleaning 45
shooting 70, 112–17, 139, 149–50
sick rooms 26–7
'sinking the beer' 137
smoking rooms 21–2
soap 58
sport 138–9, 142–52
spring cleaning 39
staff decorum 17
staff hierarchies 136–7
staff perks 12
stairs 34–5
Standen 19
Stapleton, Sir Robert 186
starching 59–60
statues 175
stockpots 82–3
stores, daily 50–1
Stourhead 175
Stowe 172, 175

Stratfield Saye 23, 144
Sudbury Hall 35
sugar, spun 100
sundials 181–2
superstition 86
Swakeley's 33–4
Syon House 19

table layouts 67–9
telephones 25- 6
Tendring Hall 103
tennis 143–4
Thompson, Sir Benjamin 23
Thompson, Robert 7, 186, 188
tipping 123-4
toast 104
topiary 178
train stations, meeting guests at
   the 119- 20
Tredegar House 162–3
Trollope, Anthony 116–17, 151

Uppark 92, 192
'Upper Ten' 136–7

valets 41, 44, 69–70, 75, 108
vegetables 85, 184
Victoria 25, 66, 122, 144

walled gardens 186–7
wallpaper 20
Walpole, Horace 152
Walton, Izaak 90
water features 179–80
Webb, Philip 19, 37
Wellbeck Abbey 26
Wellington, Duke of 23, 144
wet larders 101–2
White, Florence 103
wine 54–5, 130- 1
Woburn Abbey 23, 171
Wollaton Hall 182
Wyatt, James 35

A DAVID & CHARLES BOOK
© F&W Media International, Ltd 2012

David & Charles is an imprint of F&W Media International, Ltd
Brunel House, Forde Close, Newton Abbot, TQ12 4PU, UK

F&W Media International, Ltd is a subsidiary of F+W Media, Inc
10151 Carver Road, Suite #200, Blue Ash, OH 45242, USA

Text © Ruth Binney 2012
Layout © F&W Media International, Ltd 2012

First published in the UK in 2012

A catalogue record for this book is available from the British Library.

ISBN-13: 978-1-4463-0218-7 hardback
ISBN-10: 1-4463-0218-0 hardback

Printed in Great Britain by TJ International Ltd for:
F&W Media International, Ltd
Brunel House, Forde Close, Newton Abbot, TQ12 4PU, UK

10 9 8 7 6 5 4 3 2 1

Acquisitions Editor: Neil Baber
Assistant Editor: Hannah Kelly
Project Editor: Beverley Jollands
Junior Art Editor: Jo Lystor
Production Manager: Beverley Richardson

F+W Media publishes high quality books on a wide range of subjects.
For more great book ideas visit: www.rucraft.co.uk